Optical Transmission for the Subscriber Loop

The Artech House Optoelectronics Library

Brian Culshaw, Alan Rogers, and Henry Taylor, *Series Editors*

Acousto-Optic Signal Processing: Fundamentals and Applications, Pankaj Das

Amorphous and Microcrystalline Semiconductor Devices, Optoelectronic Devices, Jerzy Kanicki, editor

Electro-Optical Systems Performance Modeling, Gary Waldman and John Wootton

The Fiber-Optic Gyroscope, Hervé Lefèvre

Field Theory of Acousto-Optic Signal Processing Devices, Craig Scott

Highly Coherent Semiconductor Lasers, Motoichi Ohtsu

Introduction to Electro-Optical Imaging and Tracking Systems, Khalil Seyrafi and S. A. Hovanessian

Introduction to Glass Integrated Optics, S. Iraj Najafi

Optical Control of Microwave Devices, Rainee N. Simons

Optical Fiber Sensors, Volume I: Principles and Components, John Dakin and Brian Culshaw, editors

Optical Fiber Sensors, Volume II: Systems and Applicatons, Brian Culshaw and John Dakin, editors

Optical Network Theory, Yitzhak Weissman

Principles of Modern Optical Systems, Volume I, I. Andonovic and D. Uttamchandani, editors

Principles of Modern Optical Systems, Volume II, I. Andonovic and D. Uttamchandani, editors

Reliability and Degradation of LEDs and Semiconductor Lasers, Mitsuo Fukuda

Single-Mode Optical Fiber Measurements: Characterization and Sensing, Giovanni Cancellieri

UNIVERSITY OF STRATHCLYDE

30125 00463316 9

**Books are to be returned on or before
the last date below.**

22 OCT 2002

2 0 JAN 2003

1 8 MAR 2003

- 5 JUN 2003

1 2 DEC 2003

1 5 MAR 2005

LIBREX—

Optical Transmission
for the Subscriber Loop

Norio Kashima

Artech House
Boston • London

© 1993 ARTECH HOUSE, INC.
685 Canton Street
Norwood, MA 02062

All rights reserved. Printed and bound in the United States of America. No part of this book may be reproduced or utilized in any form or by any means, electronic or mechanical, including photocopying, recording, or by any information storage and retrieval system, without permission in writing from the publisher.

International Standard Book Number: 0-89006-679-5
Library of Congress Catalog Card Number:

10 9 8 7 6 5 4 3 2 1

Contents

Preface	ix

Part I Basic Technology

Chapter 1 Introduction to Optical Transmission for Subscriber Loops
- 1.1 Outline of Optical Subscriber Loops — 3
  - 1.1.1 Optical Subscriber Loop System — 4
  - 1.1.2 Optical Trunk System — 4
  - 1.1.3 Comparison — 7
  - 1.1.4 Classification of the Subscriber Loop Systems — 9
- 1.2 Network Topology for Subscriber Loops — 9
- 1.3 Multiplexing Method for Bidirectional Transmission — 11
- 1.4 Access Method for Passive Optical Network — 19
- 1.5 Transfer Mode for Digital Transmission — 26
- References — 28

Chapter 2 Fundamental Optical Devices for Transmission
- 2.1 Optical Fiber and Cable — 29
  - 2.1.1 Optical Fiber — 29
  - 2.1.2 Optical Cable — 37
- 2.2 Fiber Splice and Connector — 41
- 2.3 Laser Diode — 46
  - 2.3.1 FP-LD — 48
  - 2.3.2 DFB and DBR LDs — 53
  - 2.3.3 Modulation Characteristics — 53
  - 2.3.4 Linewidth of LD — 56
  - 2.3.5 Chirping — 57
  - 2.3.6 Equivalent Circuit — 57
- 2.4 Photodiode — 57
  - 2.4.1 Responsivity and Photocurrent — 59

		2.4.2	Wavelength Dependence of R	60
		2.4.3	Equivalent Circuit	61
		2.4.4	Noise	61
	2.5	Fiber Coupler and WDM Devices		65
		2.5.1	Fiber- and Waveguide-Type Devices	65
		2.5.2	Bulk-Type WDM Device	68
	2.6	Relationship between Optical Systems and Optical Devices		71
	References			71
Chapter 3	Digital and Analog Transmission			
	3.1	Intensity Modulation and Direct Detection		73
		3.1.1	Intensity Modulation	75
		3.1.2	Direct Detection	79
	3.2	Coherent Transmission		88
		3.2.1	Principle	89
		3.2.2	Receiver Configuration and Error Rate	94
		3.2.3	OFDM	105
		3.2.4	Application in Subscriber Loops	108
	3.3	Subcarrier Multiplexing		109
		3.3.1	SCM Systems Using IM/DD	112
		3.3.2	SCM Systems Using Coherent Technology	114
	3.4	Limitation Due to Fiber Nonlinearity		115
		3.4.1	SRS	115
		3.4.2	SBS	119
		3.4.3	Intensity-Dependent Refractive Index	122
		3.4.4	FWM	123
	References			125
Chapter 4	Optical Receivers			
	4.1	Classification and Requirements of Receiver Design		129
	4.2	Several Front-End Designs		134
		4.2.1	High-Impedance Design	134
		4.2.2	Transimpedance Design	138
		4.2.3	Low-Impedance Design	144
		4.2.4	Resonance-Type Design for Coherent Transmission	145
		4.2.5	Induction Peaking Design for High-Speed Transmission	145
	4.3	Comparison of Front-End Designs		145
	4.4	AGC and Timing Extraction		150
		4.4.1	AGC	150
		4.4.2	Timing Extraction	152
	4.5	Degradation Factors		152
	References			153

Chapter 5 Optical Amplifier
- 5.1 Outline of Optical Amplifier — 155
- 5.2 Semiconductor Laser Diode Amplifier — 160
  - 5.2.1 Gain Profile of Laser Diode Amplifier — 161
  - 5.2.2 Saturation Output Power — 164
  - 5.2.3 Polarization Dependence of Gain — 168
- 5.3 Doped-Fiber Amplifier — 172
  - 5.3.1 Configuration — 172
  - 5.3.2 Gain of EDFA — 174
  - 5.3.3 Saturation Output Power — 176
  - 5.3.4 Polarization Dependence of Gain and Coupling with a Fiber — 177
- 5.4 Noise of Optical Amplifier — 177
  - 5.4.1 Basic Noise Equation of Optical Amplifier — 177
  - 5.4.2 Noise of TWA LD Amp — 179
  - 5.4.3 Noise of FP-LD Amp — 181
  - 5.4.4 Noise of EDFA — 183
  - 5.4.5 Comparison of Optical Amplifiers — 183
- 5.5 Optical Amplifier Response — 184
- 5.6 Simultaneous Amplification for WDM and OFDM — 186
- References — 189

PART II System Examples and Optical Devices

Chapter 6 Bidirectional Systems Using TCM
- 6.1 TCM System Configuration and Examples — 195
- 6.2 Laser Diode as a Photodetector in TCM — 197
  - 6.2.1 LD Transceiver in TCM — 197
  - 6.2.2 Photodetection Properties of a Laser Diode — 200
  - 6.2.3 Performance of LD Transceiver — 209
- 6.3 Laser Diode as an Attenuator in TCM — 211
  - 6.3.1 LATT Configuration — 211
  - 6.3.2 Basic LATT Properties — 213
  - 6.3.3 Dynamic Range of Receiver with LATT — 215
- 6.4 Laser Diode as a Preamplifier in TCM — 218
  - 6.4.1 Receiver Configuration — 218
  - 6.4.2 Transmission Experiment Using a Conventional Laser Module — 218
  - 6.4.3 Receiver Sensitivity — 222
- References — 227

Chapter 7 Optical Transmission Systems Using WDM
- 7.1 WDM for Upgrading Method — 229

		7.1.1 Upgrading Methods	229
		7.1.2 WDM Devices	233
		7.1.3 Design Consideration of WDM Systems for Upgrading	233
	7.2	WDM-based Multichannel Systems	236
	7.3	WDM- and SCM-based Multichannel Systems	240
	References		245

Chapter 8 Optical Transmission Systems Using Coherent Technology
    8.1 Tunable Laser Diode   247
        8.1.1 Multielectrode Tunable Laser Diode   248
        8.1.2 External Cavity Tunable Laser Diode   251
        8.1.3 Tuning Characteristics of Tunable Laser Diode   251
    8.2 Optical Multi/Demultiplexer for OFDM   253
        8.2.1 Mach-Zehnder Interferometer Type   259
        8.2.2 Fabry-Perot Interferometer Type   263
        8.2.3 Laser Diode Filter Type   264
        8.2.4 Other Types   264
    8.3 Passive Optical Network Using OFDM   265
        8.3.1 Network Configuration   265
        8.3.2 Experiments   273
    8.4 Multichannel System Using Coherent Technology   278
        8.4.1 Distribution Systems   278
        8.4.2 Bidirectional Systems   280
    References   285

Chapter 9 SCM Systems
    9.1 Multiple-Access Network   289
    9.2 Multichannel Video Distribution System   291
    9.3 Enhanced Multichannel Video Distribution System Using EDFA   293
    References   297

Chapter 10 Future Subscriber Systems
    10.1 Photonic Integrated Circuits   299
    10.2 Broadband Systems Using Optical Signal Processing   299
    10.3 Multiwavelength System Using Multiple-Access Method   303
    References   303

Appendix A List of Acronyms   305

Index   309

Preface

Optical transmission systems for trunk lines are used worldwide. Optical transmission systems using optical fibers will penetrate deeply into the subscriber loops in the future. Although the basic technologies are common to both trunk and subscriber systems, optical transmission for subscriber loops must be designed by considering its characteristics and peculiarities. The bit rate is low when compared to the trunk line, so systems that are more economical must be deployed. The sharing systems and economical receiver designs have been investigated for this purpose. Multiple-access systems and multichannel video distribution systems may be limited to subscriber loop systems.

    This book deals with optical transmission technologies using single-mode optical fibers for subscriber loops. Although the technologies described in this book are intended for systems using single-mode fibers, many of them are applicable to systems based on multimode optical fibers. Optical subscriber loop systems using single-mode fibers are not widely penetrated at present; therefore, the many technologies explained in this book are not practical now, but might be used widely in the future. This book is intended for system designers, technical managers, students, and manufacturers who are interested in optical subscriber loop systems. It is also useful for the researcher who wants a summary of this area.

    This book is divided into two parts: Part I deals with the basic technology and Part II with optical subscriber system examples and optical devices used in the systems. Chapter 1 contains introductory material for optical subscriber loop systems. The characteristics of subscriber loops are explained by comparing them to those of a trunk system. This chapter also contains the explanation of network topology, the multiplexing method, the access method, and the transfer mode. Chapter 2 contains an explanation of fundamental optical devices. Chapters 3 and 4 describe optical fiber transmission. Chapter 3 treats digital and analog transmission, as well as coherent transmission and the ordinary intensity modulation and direct detection schemes. Chapter 4 explains the design of the optical receiver, including active feedback design and optical feedback design. Chapter 5 describes the semiconductor laser diode amplifier and the doped fiber amplifier. It also explains simultaneous amplification for multiple-access systems and multichannel video distribution systems. Chapters 6 through 10 treat system examples that are mainly discussed in laboratories. Chapter 6 contains bidirectional systems using time-compression multiplexing, and it explains several usages of a laser diode, (e.g., as a photodetector, attenuator, preamplifier, and light source). Chapters 7 and 8 describe systems using wavelength-division multiplexing and coherent technology. These technologies are explained for multichannel systems. Passive optical networks using optical frequency-division multiplexing are also explained. Chapter 9 describes subcarrier multiplexing systems and gives examples of multiaccess and multichannel video distribution systems. Chapter 10 contains the proposed future subscriber systems, which are only a dream at present but may be a reality in the near future.

This book covers a wide range of technologies, such as optical devices (an optical fiber, a laser diode, wavelength-division multiplexing devices, etc.), digital and analog transmission, and optical amplification. Therefore, the same notations have different meanings. For example, α is used for the α-parameter, which represents the refractive index profile for fibers. The same notation α is also used for the linewidth enhancement factor of a laser diode. To avoid confusion, it is possible to adopt a unique notation system in this book (e.g., changing the α-parameter to β-parameter in fibers, but not changing it in lasers). However, I have used common notations because they are convenient for the reader. I spoke with fiber people about the refractive index profile using the α-parameter and discussed the linewidth using α with laser people. Since the notation α is well known to them, it is better to use α for these notations. To avoid confusing the reader, throughout the book I have defined the notations used in equations.

I wish to thank my colleagues at Nippon Telegraph and Telephone (NTT) for helping me in my research. I also thank Mr. Mark Walsh for inviting me to write this book and for his encouragement.

PART I
Basic Technology

Chapter 1
Introduction to Optical Transmission for Subscriber Loops

Subscriber loop systems have different characteristics compared to trunk systems. In this chapter, the different characteristics and classifications of subscriber loop systems are explained, as well as the basic concepts and technologies of network topology, multiplexing methods, access methods, and transfer modes.

1.1 OUTLINE OF OPTICAL SUBSCRIBER LOOPS

Many telecommunication services, such as telephone, facsimile, and data transmission between computers, are provided to our homes and companies. We can receive not only domestic but also international services through satellites in space or cables under the sea. Telecommunication is categorized into international and domestic communication. Domestic telecommunication is composed of trunk systems and subscriber loop systems. Trunk systems connect *telephone offices* (TO), and subscriber loop systems connect TOs with subscribers (users).

At an early stage, optical transmission using free space and gas laser was investigated. Since the development of low-loss optical fiber, optical transmission using optical-fiber cable has been investigated and widely used. Historically, optical trunk transmission systems were first investigated and then introduced into real use. In early systems, 0.8-μm wavelength light and step-index multimode fibers were used. The optical light source was a *light-emitting diode* (LED) or a *laser diode* (LD). Next, 1.3-μm wavelength light and graded-index multimode or single-mode fibers were used. The optical light source was a laser diode. Today's trunk systems generally use 1.3- or 1.5-μm wavelength light, single-mode fibers, and *distributed-feedback laser diodes* (DFB-LD).[1] Construction of optical subscriber transmission systems has been the next target for researchers. In the early investigation stage of

---
[1] Fibers and lasers are explained in Chapter 2.

the optical subscriber loop system, systems using graded-index multimode fiber were taken into consideration. At present, single-mode optical fiber systems are being investigated and are regarded as preferable systems.

1.1.1 Optical Subscriber Loop System

An example of an optical subscriber loop system using optical fiber cables is shown in Figure 1.1. In this figure, two types of systems are shown:
1. Subscribers in a user building or a home are directly connected to the TO through optical cables. Information is carried directly by optical transmission systems.
2. Subscribers are connected to the TO through optical and metallic cables. Information conveyed by many metallic cables is multiplexed in a *remote terminal* (RT) and converted into a light by an *optical/electrical* (O/E) converter (e.g., using a laser diode). The converted and multiplexed information is transmitted by one fiber. In an RT, there are multiplexers and demultiplexers, which are made by electrically active devices such as *large-scale integrated circuits* (LSI). In this case, an optical transmission system is between a TO and RT.

The typical transmission length for a subscriber loop system is within 10 km. In this case, when a TO is located in the center, subscribers are scattered in a circle with a 5-km radius.

1.1.2 Optical Trunk System

An example of an optical trunk system using optical-fiber cables is shown in Figure 1.2. Information from subscribers in a TO is selected and gathered by an exchanger and is multiplexed. The multiplexed information is transmitted between TOs at a higher bit rate (in the case of digital transmission) when using optical-fiber cables. At present, digital transmission using single-mode fibers is commonly used. For long-distance transmission, repeaters are used. In a repeater, received digitally modulated light signals are converted to electric signals by a *photodiode* (PD), and the converted signals are reshaped electrically through the so-called 3R functions (retiming, reshaping, and regenerating). The electric signals are again converted to light signals by a laser diode. A typical repeater interval is from 50 to 100 km.

At present, 3R function and amplification are performed by electric devices. Optical signal processing without converting to electric signal forms is being investigated for higher speed transmission. Soliton transmission using fiber nonlinearity is also being investigated for future trunk transmission. In soliton transmission, a solitary pulse can transmit over long distances without narrowing or broadening the pulse width only by using optical amplifiers. When these technologies are used, an

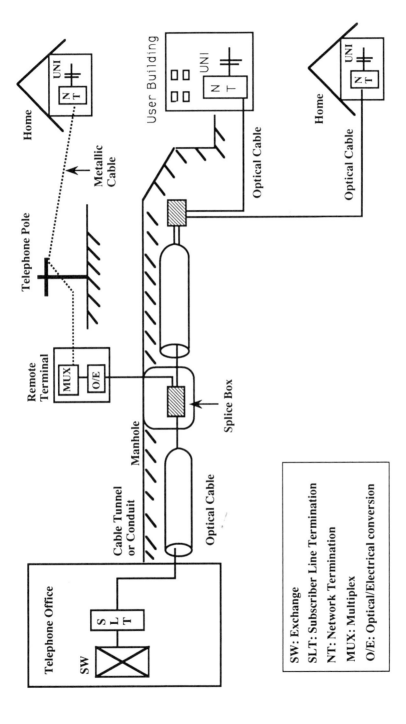

Figure 1.1 Optical subscriber loop systems.

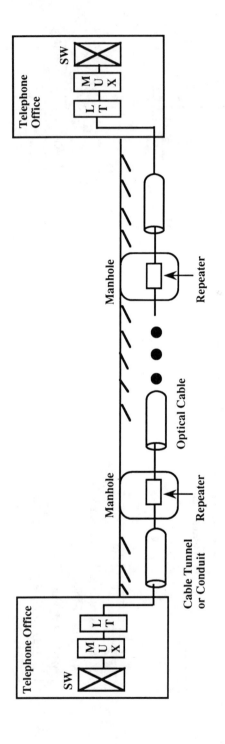

Figure 1.2 Optical trunk system.

optical trunk system is realized without conversion to electric signals. These technologies may also influence future subscriber systems.

1.1.3 Comparison

A comparison of subscriber transmission systems and trunk transmission systems is displayed in Table 1.1. In some real systems, system characteristics are not as distinct as those shown in the table. Values used in a table are typical at present. Since subscriber systems are constructed according to direct user demands, and the user demands are not uniform, network configuration is generally complex. On the other hand, network configuration of a trunk system is very simple because a trunk system connects predetermined TOs. In trunk systems, line bit rate is high as a result of multiplexing. A trunk system is shared by many users so that the system can easily become economical. Cost reduction is very severe in subscriber loop systems. When the number of subscribers using optical transmission systems increases in the future to the number of present telephone subscribers, the number of optical subscriber systems will be very large. For cost reduction and coping with the large number of systems, mass production technologies must be developed, such as simple equipment and component configuration. Transmission equipment for subscriber systems must be small because the equipment is placed on user sites. Small equipment has a small heat dissipation area. To realize the small equipment, low power consumption devices must be developed and used. For example, to accomplish a given function, *complementary metal oxide semiconductor* (CMOS) devices are preferable to bipolar devices.

Table 1.1
Comparison of Subscriber Transmission Systems to Trunk Systems

Items	*Subscriber Systems*	*Trunk Systems*
Network configuration	Complex	Simple
Line bitrate	Moderate (1.5Mb/s–600 Mb/s)*	High (100 Mb/s–10 Gb/s)*
Transmission length	Short (0–10 Km)*	Long (repeater interval: 50 Km–100 km)*
Number of systems	Very large	Not so large
Equipment environment	Poor	Good
Cost	Severe	Not so severe (shared by many users)

* Typical value

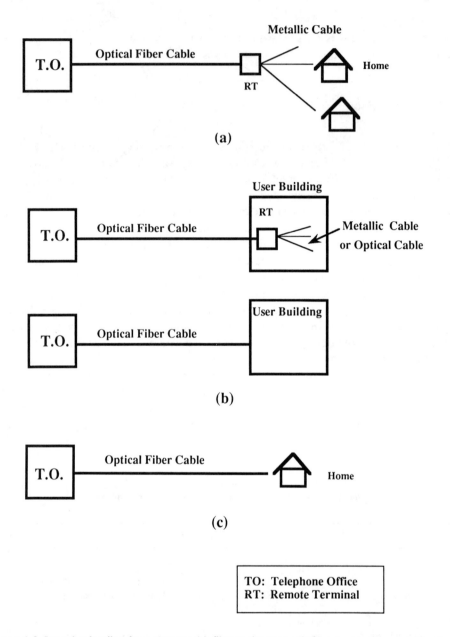

Figure 1.3 Several subscriber loop systems: (a) fiber-to-the-curve; (b)fiber-to-the-office; (c) fiber-to-the-home.

1.1.4 Classification of the Subscriber Loop Systems

The classifications of subscriber loop systems are named after optical-fiber cable penetration sites, as shown in Figure 1.3: *fiber-to-the-curve* (FTTC), *fiber-to-the-office* (FTTO), and *fiber-to-the-home* (FTTH) [1]. In FTTC, multiplexed signals are transmitted between the TO and the RT through optical cables. FTTO is the optical subscriber system for business use, and two configurations of this may be possible, as shown in the figure. In FTTH, a house is directly connected to the TO only using optical cables. For FTTH, several issues must be clarified and solved. One of these is power feeding, which is not treated in this book. Since a metallic cable can transmit a dc current, a traditional telephone at home is powered by a TO. A metallic cable can be used as a transmission medium both for signals and electric power. In the case of optical fiber, only light can be transmitted. When FTTH is realized, broadband services, such as video distribution and video conferencing, can be provided to the home through optical cables.

1.2 NETWORK TOPOLOGY FOR SUBSCRIBER LOOPS

When compared to metallic cables, optical-fiber cables have splendid characteristics, such as: wide bandwidth (that is possible for multiplexing in an electrical domain); the possibility of wavelength division multiplexing; very low loss. Considering these characteristics, many network topologies are being proposed and considered for optical subscriber loops and are shown in Figure 1.4.

Single star (SS) is the network that is directly connected between the TO and the user without optical splitters or electric multiplexers. This is very simple and the components used between optical cables in this network are only splicing devices or optical connectors. Most of the traditional telephone network using metallic cables uses this SS topology. *Passive double star* (PDS), *passive triple star* (PTS), and, more generally, passive multiple star belong to the *passive optical network* (PON) [2]. PON is the general concept of these networks, which are multiplexed by using passive, not active, components. The active components mean that the components are using electric power. Examples of active components are multiplexers composed of electric circuits and *wavelength-division multiplex* (WDM) components using *acousto-optical* (AO) devices. WDM components using a grating or fused-fiber coupler or waveguide-type interferometer and an optical splitter using a fused-fiber coupler are examples of passive components. *Active doubler star* (ADS) uses active components for multiplexing. This electric multiplexer is located in the RT. From the RT to the users, both types of cables, optical or metallic, are possibly used. Active WDM components, such as those using AO devices instead of an electric multiplexer, may be replaced in ADS topology, although this is not shown in Figure 1.4. *Hybrid star* is a mixture of passive and active star, and many modified topologies can be considered. *Bus* is used in traditional electric wiring, such as coaxial

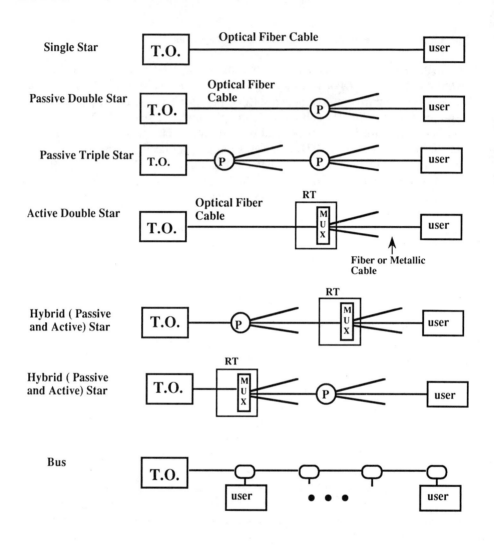

Figure 1.4 Network topology.

cable television networks, *local-area networks* (LAN), and connections in computer hardware. Bus topology is similar to electric bus topology. Dropping is made by either passive or active components. Examples of passive components for dropping are fused-fiber couplers and passive WDM devices. One example of an active component for dropping is made from the combination of O/E and E/O devices, which is similar to a repeater in trunk systems. Another example of dropping is made from a photodiode that detects leakage light from a fiber. This dropping method is realized without cutting a fiber.

The selection of network topologies may be affected by several factors:
1. Whether to provide broadband or narrowband service;
2. Whether to provide bidirectional (communication) or unidirectional (broadcast) service;
3. Whether the transmission distance from the TO to subscribers is short or long;
4. Subscriber density;
5. Upgradability from narrowband service to broadband service transmission;
6. Ease of operation, including network management.

1.3 MULTIPLEXING METHOD FOR BIDIRECTIONAL TRANSMISSION

Bidirectional transmission is necessary for communication. Figure 1.5 lists several ways of realizing bidirectional transmission. These are divided into two categories: one fiber or two fibers. The optical configurations of these methods are shown in Figure 1.6. They are *space-division multiplexing* (SDM), WDM, *directional division multiplexing* (DDM), *time-compression multiplexing* (TCM), *code-division multiplexing* (CDM), and *subcarrier multiplexing* (SCM). In SDM, two fibers are used for downstream and upstream transmission. This is the simplest among them, but the number of fibers is twice that of the others, so the transmission line cost is the highest. In WDM, upstream and downstream transmissions are made over a single fiber at different wavelengths. They are multiplexed and demultiplexed using WDM devices. Two types of lasers and WDM devices must be used, and this results in a

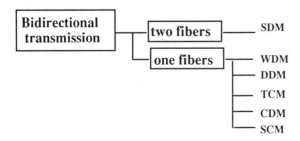

Figure 1.5 Classification of bidirectional transmission schemes.

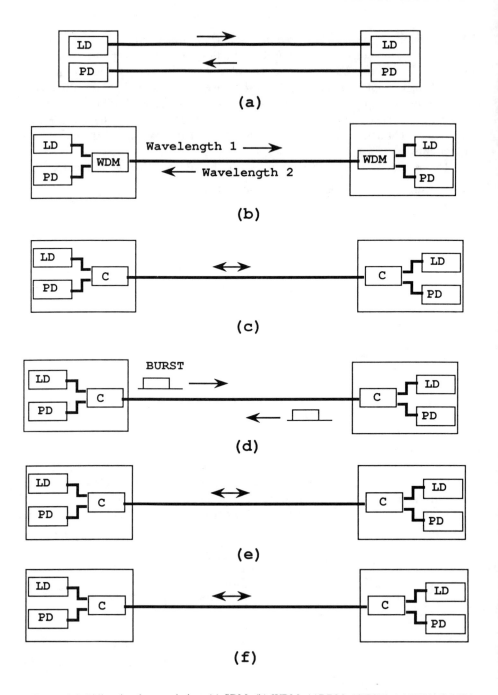

Figure 1.6 Bidirectional transmission: (a) SDM; (b) WDM; (c)DDM; (d)TCM; (e)CDM; (f)SCM.

relatively high-cost system. In DDM, LDs and PDs are combined by using two couplers, and the same wavelength is used for both streams. In this system, a countermeasure against reflection light must be considered, as is explained later. Construction of TCM, CDM, and SCM is similar to that of DDM. These methods use the same wavelength for both streams. They differ in the handling of the electric signal. In TCM, data are transmitted in bursts (ping-pong transmission), not continuously. Therefore, reflection light has no effect. Memory must be used to convert burst data to continuous data, and higher bit rate transmission is required due to TCM. Therefore, TCM is suitable for relatively low-bit-rate transmission. In CDM, upstream and downstream transmissions are modulated by different codes. The effect of reflection light can be eliminated by using different code multiplexing because the recovery of signals is accomplished through correlation, and different code has less or no correlation. Through the modulation process of CDM, signals result in a higher bit rate. In SCM, upstream and downstream signals are modulated by a different frequency subcarrier. The effect of reflection light can be eliminated by using electric filters. Since SCM is analog transmission, severe linearity is required for some applications.

Some further explanations of WDM, DDM, TCM, CDM, and SCM are illustrated in Figures 1.7 to 1.11. Figure 1.7 shows the crosstalk in WDM. In the case of nonideal isolation (backward isolation) of a WDM device, part of sending light is back to a photodiode. When a photodiode has the sensitivity for both wavelengths, the backward light causes crosstalk. For example, 1.3 and 1.5 μm are used for both wavelengths, and InGaAs pin photodiodes are used. InGaAs pin photodiodes can detect 1.3- and 1.5-μm wavelengths.

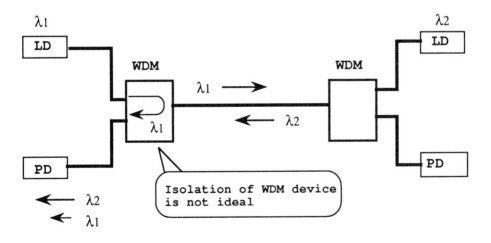

Figure 1.7 Crosstalk in WDM transmission.

The optical reflection in DDM transmission is shown in Figure 1.8. The reflection of optical devices such as optical connectors takes place. Both reflected and transmitted lights are detected by a photodiode and cause the degradation of receiver sensitivities. The reflection of an ordinary optical connector against incoming light ranges from -15 to -35 dB in optical level. Here, a -15 dB reflection in optical power means -15 dBm optical power of reflected light when sending optical power is 0 dBm. Some connectors have a lower reflection (e.g., under -45 dB). In an ordinary optical transmission line, many connectors are used. The reflected light level increases in this case.

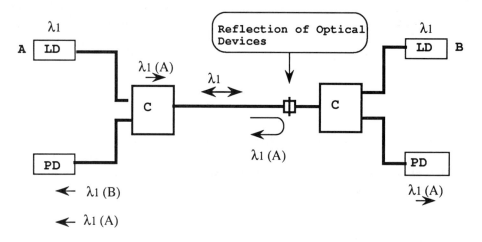

Figure 1.8 Optical reflection in DDM transmission.

The TCM system is shown in Figure 1.9 [3]. The timing chart in this figure is shown for one burst cycle. T_D, T_G, and T_{inf} are delay time, guard time, and information time, respectively. The delay time T_D corresponds to the transmission time of optical signals in fiber cables. Typically, $T_D = 5L$ (ns), where L is the transmission distance in meters. Guard time T_G corresponds to the switching time between the receiving and transmission modes. Burst cycle time T_B is equal to $2(T_{inf} + T_D + T_G)$, where the time needed for housekeeping bits is included in T_{inf}. The time for preamble bits and burst frame synchronous bits is also included in T_{inf}. Line bit rate B in the TCM system depends on the transmission distance, burst cycle, line code, and guard time. For the *nonreturn to zero* (NRZ) line code, the following equations hold from Figure 1.9.

$$T_B \geq 2\left(T_D + T_G + \frac{I_{inf}}{B}\right) \quad (1.1)$$

$$I_{inf} = B_l T_B \quad (1.2)$$

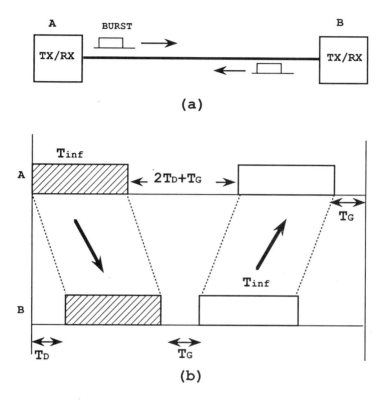

Figure 1.9 TCM transmission systems: (a) configuration; (b) timing.

where I_{inf} is the information bits and B_I is the bit rate of source information (not burst, but continuous bit stream). Using (1.1) and (1.2), the necessary line bit rate B in the TCM system is obtained as follows.

$$B \geq \frac{2B_I T_B}{[T_B - 2(T_D + T_G)]} \tag{1.3}$$

From (1.3), the required line bit rate B in the TCM system is more than two times the original source bit rate B_I. For the *coded mark inversion* (CMI) code, B is doubled. The memory capacitance for time compression may be different according to the type of memory Typical memory capacitance is the following, considering the reading and writing period.

$$M = 2 I_{inf} \tag{1.4}$$

For a numerical example, $T_B = 1$ ms, $L = 10$ km ($T_D = 50$ μs), $T_G = 5$ μs, and

$B_I = 1.5$ Mb/s are assumed; then B is about 3.4 Mb/s in the case of NRZ code and 3 kb of memory is necessary.

Figure 1.10 shows the principle of CDM [4]. Two different codes are used for both streams. Input electric signals are multiplied by the unique code for each stream and then converted to optical signals. Here, we use notations $S_1(t)$, $C_A(t)$, $X(t)$ and t, which represent an input signal, a code for CDM, output signals from the CDM modulator, and time for upstream transmission, respectively. $S_2(t)$, $C_B(t)$, and $Y(t)$ represent an input signal, a code for the CDM, and output signals from the CDM modulators for downstream transmission, respectively. For CDM modulation, the following equations hold using these notations.

For upstream transmission,

$$X(t) = S_1(t) \times C_A(t) \qquad (1.5)$$

and for downstream transmission,

$$Y(t) = S_2(t) \times C_B(t) \qquad (1.6)$$

In CDM, bit rates of codes $C_A(t)$ and $C_B(t)$ are much higher than those of input signals $S_1(t)$ and $S_2(t)$. These are transmitted by lasers and fibers both ways. Recovery of $S_1(t)$ for upstream transmission is made by the multiplication of $C_A(t)$ with (1.5):

$$X(t) \times C_A(t) = S_1(t) \times C_A(t) \times C_A(t) = S_1(t) \qquad (1.7)$$

In (1.7), $C_A(t) \times C_A(t)$ is equal to 1 because $C_A(t)$ takes the value of $+1$ or -1. The reflected light is also detected generally by a photodiode. It is expressed by

$$Y(t) \times C_A(t) = S_2(t) \times C_B(t) \times C_A(t) \qquad (1.8)$$

The product $C_B(t) \times C_A(t)$ is like a random signal (codes $C_A(t)$ and $C_B(t)$ are selected for having this characteristic) so that $Y(t) \times C_A(t)$ is like a random signal with a broad bandwidth. In practice, there is less or no intersymbol interference with the recovering process of $S_1(t)$.

Figure 1.11 shows the principle of SCM. In SCM, upstream and downstream signals are modulated by different frequency subcarriers. Original signals are upconverted through the multiplication of the unique subcarrier and a *band-pass filter* (BPF). As shown in Figure 1.11, upstream and downstream signals are placed in frequency domain so as not to overlap by selecting the appropriate subcarrier frequencies. It is possible to set $f_{c1} = 0$ as the baseband transmission for one stream. The effect of the reflection light can be eliminated by using electric filters.

A comparison of SDM, WDM, DDM, TCM, CDM, and SCM is made in Table 1.2. In DDM, optical devices are complex to minimize the effect of reflected light.

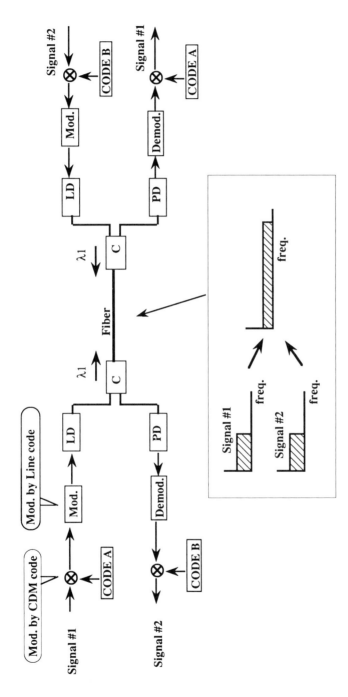

Figure 1.10 Principle of CDM.

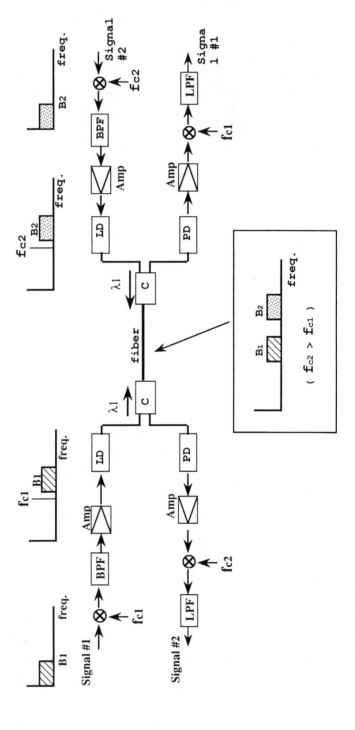

Figure 1.11 Principle of SCM.

Table 1.2
Comparison of Bidirectional Transmission Schema

Items	SDM	WDM	DDM	TCM	CDM	SCM
Number of fibers	2	1	1	1	1	1
Line bit rate	B	B	B	~2B	≫2B	Analog
Number of wavelengths	1	2	1	1	1	1
Complexity of optical device	○	●	●*	○	○	○
Complexity of electrical circuit	○	○	○	●	●	●

Note: ○ = simple; ● = complex; B = bit rate.

* Complex, due to antireflection.

An antireflection coating may be applied to the device facets, and refractive index matching such as physical or optical contact may be required for optical connectors. In TCM, complex electric circuits are necessary for burst synchronization (TCM synchronization). For CDM and SCM, complex electric circuits are also needed for multiplying a code or a subcarrier.

1.4 ACCESS METHOD FOR PASSIVE OPTICAL NETWORK

Several access methods for restricted communication resources have been proposed and investigated. Multiple-access methods are used for satellite communication, as shown in Figure 1.12. One satellite is used by many satellite offices. There are *frequency-division multiple access* (FDMA), *time-division multiple access* (TDMA), *code-division multiple access* (CDMA), and *space-division multiple access* (SDMA). In FDMA, a unique frequency is assigned to the satellite office and signals are distinguished by this frequency. In TDMA, time is divided and some of the divided time is assigned to the satellite office. In TDMA, each satellite office sends and receives signals in bursts, not continuously, using the same frequency. Signals are distinguished by the position in time domain or the label used for *identification* (ID) in signals. In CDMA, a unique code is assigned to the satellite office, and the signal spectrum spreads out in the frequency domain. Signals are detected using the unique code. In SDMA, antenna beams are used in a satellite and each beam is assigned to the satellite office.

Similar multiple-access methods may be used for optical subscriber loop systems. It is being proposed that PON shares optical fibers and equipment in a TO. Three applications of PON using a *star coupler* (SC) are shown in Figure 1.13. They differ according to the location of the SC. Multiple-access methods must be applied

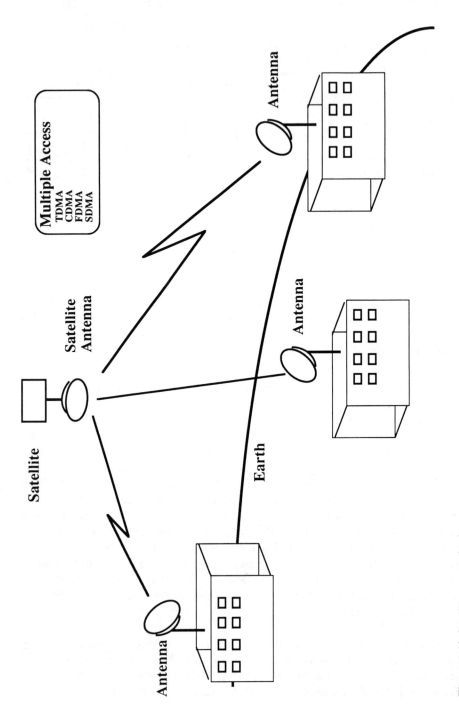

Figure 1.12 Satellite communication.

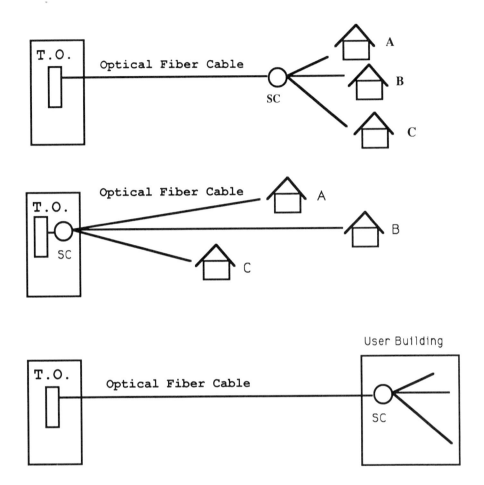

Figure 1.13 Passive optical network.

Table 1.3
Multiple Access Schema in PON

	Resources for Sharing	*Multiple-Access Schemes*
Optics	Wavelength	WDMA (OFDMA)
Electronics	Time	TDMA
	Code	CDMA
	Frequency	SCMA

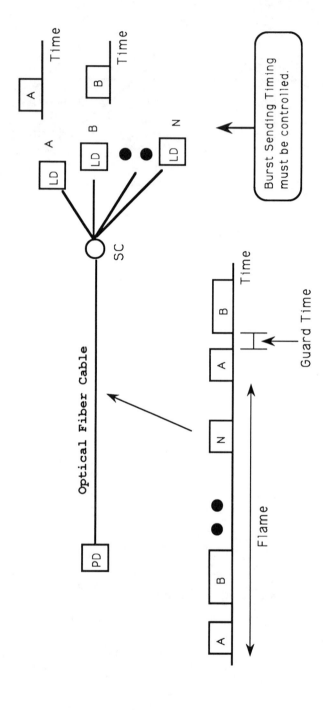

Figure 1.14 TDMA for PON.

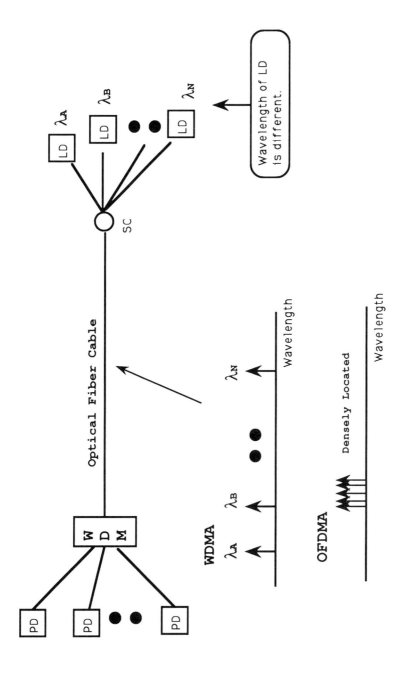

Figure 1.15 WDMA (OFDMA) for PON.

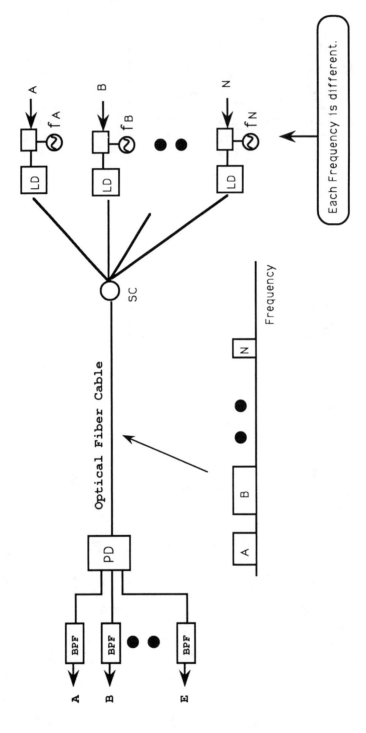

Figure 1.16 SCMA for PON.

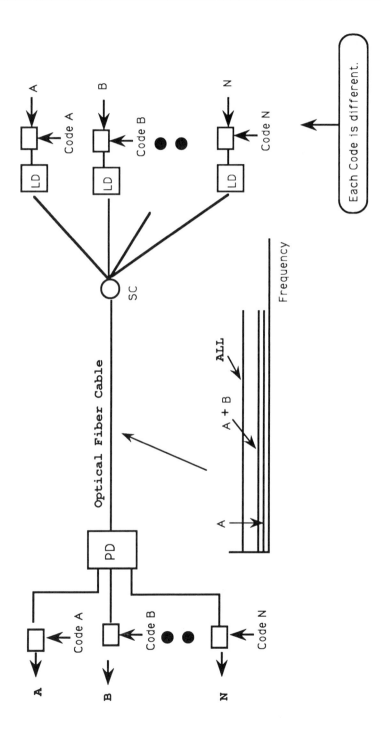

Figure 1.17 CDMA for PON.

for upstream signals. Multiple-access methods in PON, which are analogous to multiple-access methods in satellite communication, are listed in Table 1.3. They are classified according to resources for sharing. *Optical frequency-division multiple access* (OFDMA) is the special case of *wavelength-division multiple access* (WDMA) with very narrow wavelength spacing. Explanations of TDMA, OFDMA, *subcarrier multiple access* (SCMA), and CDMA are illustrated in Figures 1.14 to 1.17. In TDMA, the burst sending timing at each piece of equipment must be controlled so as not to overlap in time domain. This is done by using the control signals from a TO (it is assumed that a PD is placed in a TO). Since the control signals are needed, the TDMA scheme is a rather complex access method. In WDMA, the wavelengths of all lasers must be different and are complex or expensive for optical devices. However, the access method is very simple because each wavelength is considered to be an independent transmission medium when wavelengths are ideally separated by a WDM device (a WDM device is shown as WDM in Figure 1.15). In SCMA, the frequency of each subcarrier is different so as not to overlap in the frequency domain. The access method is simple because each divided bandwidth is considered to be an independent transmission medium. In CDMA, each code is different. The access method is simple because no control signals are necessary. Table 1.4 contains a comparison of the multiple-access schemes.

Table 1.4
Comparison of Multiple Access Schema

Items	TDMA	WDMA	SCMA	CDMA
Access method	●	○	○	○
Line bit rate	N B	B	Analog **	≫N B
Number of wavelengths	1	N	1	1
Complexity of optical device	○	●	○	○
Complexity of electric circuit	●	○	●	●

Note: ○ = simple; ● = complex; N = number of multiple access; B = bit rate.

** Required bandwidth is N times.

1.5 TRANSFER MODE FOR DIGITAL TRANSMISSION

There are transfer modes for sending digital signals. They are *synchronous transfer mode* (STM), *asynchronous transfer mode* (ATM), and packet mode. Here, packet mode is considered to be X.25 packet-mode switching, defined by the *Consultative*

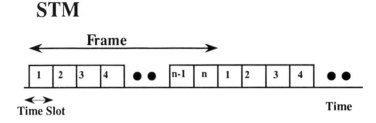

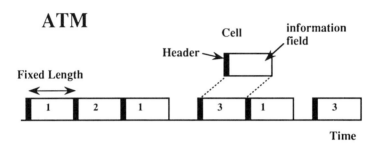

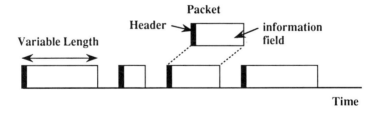

Figure 1.18 Transfer mode for digital transmission.

Committee in International Telegraphy and Telephony (CCITT). Traditionally, STM and packet mode have been used.

STM has been used as circuit-mode switching and time-division multiplexing. The fundamental unit for switching in STM is a time slot. Information from a certain channel is decomposed into time slots, which are placed in the determined time position in a frame, as shown in Figure 1.18. A frame is 125 μs in general. The bit rate for a service is a multiple of 64 kb/s. A channel in STM is identified by the position of its time slots. In the case of there being no information in a certain channel, the determined time slots used for this channel must not be diverted to other channels. Therefore, STM is not generally efficient.

Packet mode has been used as packet-mode switching. Information is decomposed into a packet, which is composed of a header and an information field, as shown in Figure 1.18. A channel in packet mode is identified by the label written in a header. The length of a packet is variable and a packet is generated asynchronously when information is being sent. This allows packet mode to be possible for any bit rate services and to be efficient. In packet mode, protocols are complex and processed by software. Therefore, packet mode is difficult for high-bit-rate use.

Among these modes, STM may be the best solution for fixed-rate services. The *broadband integrated services digital network* (BISDN) has been studied for future telecommunication. In BISDN, various services ranging from low bit rate to high bit rate must be transmitted efficiently. STM is not efficient for this purpose and packet mode is not suitable because of the high bit rate in BISDN. ATM is devised for efficient high-bit-rate transmission. Information is decomposed into an ATM cell with a header and information field as shown in Figure 1.18. The length of a cell is fixed and the cell number is variable for efficient transmission of various-bit-rate services. It has been determined by CCITT that the ATM cell is 53 bytes long with a 5-byte header. A channel in ATM is identified by the label written in a header. In ATM, protocols are simple so that they can be processed by high-speed hardware. ATM is regarded as a suitable mode for BISDN [5].

REFERENCES

[1] "The 21st Century Subscriber Loop," *IEEE Communication Magazine*, special issue, Vol. 29, No. 3, March 1991.
[2] Faulkner, D. W., D. B. Payne, J. R. Stern, and J. W. Ballance, "Optical Networks for Local Loop Applications," *IEEE J. of Lightwave Technol.*, Vol. 7, No. 11, 1989, pp. 1741–1751.
[3] Kashima, N., "Time Compression Multiplex Transmission System Using a 1.3-μm Semiconductor Laser as a Transmitter and a Receiver," *IEEE Trans. Comm.*, Vol. 40, No. 3, 1992, pp. 584–590.
[4] Dixon, R. C., *Spread Spectrum Systems*, John Wiley, 1976.
[5] Minzer, S. E., "Broadband ISDN and Asynchronous Transfer Mode (ATM)," *IEEE Communication Magazine*, Vol. 27, No. 9, September 1989.

Chapter 2
Fundamental Optical Devices for Transmission

The key devices for optical transmission are a laser diode and low-loss optical fiber. A ruby laser was developed by T. H. Maiman in 1960 [1]. *Continuous wave* (CW) lasing of a laser diode at room temperature was achieved by I. Hayashi and his colleague in 1970 [2]. Lasers with 0.8-, 1.3-, and 1.5-μm wavelengths are available at present for optical transmission. Low-loss optical fiber was anticipated by K. C. Kao in 1966 [3], and fused silica fiber with 20-dB/km loss was developed in 1970 [4]. Currently, fiber loss has decreased to below 0.2 dB/km at a wavelength of 1.55 μm. It is interesting that both CW lasing at room temperature and low-loss fiber were realized in 1970.

In this chapter, fundamental optical fiber, cable, splice, connector, laser diode, photodiode, fiber coupler, and WDM devices are explained. These are important components for optical transmission.

2.1 OPTICAL FIBER AND CABLE

2.1.1 Optical Fiber

Light from an optical light source, such as a laser diode, propagates in optical fibers. Optical fiber is made from fused silica glass (SiO_2). The refractive index of SiO_2 is about 1.45 to 1.46 near the 1-μm wavelength, and its value decreases slightly for longer wavelengths. The basic phenomenon of guiding light in a fiber is the total internal reflection, and light is confined in higher refractive-index materials. For guiding light, a fiber cross section is not uniformly made. It has a slightly higher refractive index part in the center, called *core*. Core is surrounded by cladding, as shown in Figure 2.1. To form a core and cladding structure, some dopants are used. GeO_2 is used as a dopant for increasing refractive index and fluoride (F) is for decreasing it. One example of fiber manufacturing is the combination of GeO_2 + SiO_2 for core and SiO_2 for cladding. By controlling the volume of GeO_2, the refractive index profile of core, $n_1(r)$, can be arbitrarily made (r is the radial coordinate). Another example of fiber manufacturing is the combination of SiO_2 for core and SiO_2 + F for cladding. In this case, the refractive index profile of core, $n_1(r)$, is uniform. Generally, a fiber with uniform $n_1(r)$ (i.e., $n_1(r)$ = constant) is called *step-index* (SI) fiber. The fiber mother rod is made using several vapor deposition methods. Typical

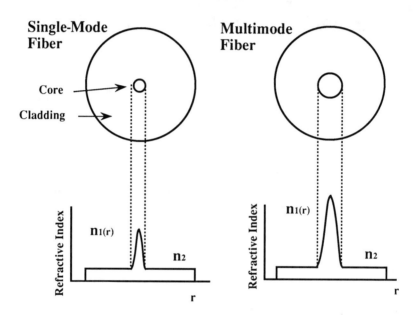

Figure 2.1 Structure of optical fiber.

methods are the *outer vapor deposition* (OVD) method, the *modified chemical vapor deposition* (MCVD) method, and the *vapor-phase axial deposition* (VAD) method. Fibers are made by drawing from the melted fiber mother rod. The melting temperature for silica fibers is about 2000°C.

Fibers with a small refractive index difference between core and cladding are usually called *weakly guiding fibers* [5]. All fibers currently used for optical transmission are weakly guiding fibers. The refractive index profile for weakly guiding fibers is usually expressed using the α parameter as follows:

$$n_1(r) = n_0 [1 - 2\Delta(r/a)^\alpha]^{1/2} \tag{2.1}$$

where

$$\Delta = (n_0 - n_2)/n_0 \tag{2.2}$$

where a is the core radius ($2a$ is the core diameter) and n_2 is the refractive index of cladding. Since the fibers being considered are weakly guiding fibers, $\Delta \ll 1$ holds. For SI fiber, $\alpha \to \infty$ (α tends to infinity) in (2.1). When we deal with fibers from the electromagnetic viewpoint, the following normalized frequency parameter V is very useful.

$$V = 2\pi a(n_1^2 - n_2^2)^{1/2}/\lambda \tag{2.3}$$

$$\approx 2\pi a n_0 (2\Delta)^{1/2}/\lambda \tag{2.4}$$

where λ is a wavelength in free space. Equation (2.4) holds because of weakly guiding fibers.

For SI fibers, only one mode can propagate in a fiber when $V < 2.405$. The value 2.405 is called the cutoff value V_C. The mode is the HE_{11} mode, or LP_{01} mode as a weakly guiding fiber. For the general α value, V_C is different from 2.405 and is larger. Fibers having one propagating mode are named as a single-mode or monomode fiber. In realizing a single-mode fiber, there are many combinations of a and Δ, determined by considering bending loss, splice loss, and manufacturing conditions. The typical values of commonly used single-mode fibers are shown in Figure 2.1. The mode field diameter is a better parameter than a and Δ for a single-mode fiber because the parameter is closely related to bending loss and splice loss [6]. Strictly speaking, the shape of the fundamental mode is different from the Gaussian shape. However, the Gaussian shape is a good approximation for the fundamental mode shape in a single-mode fiber [7]. In a single-mode fiber, two orthogonal polarized modes can generally propagate. Single-mode fibers that maintain polarizations (called *polarization holding fibers* or *polarization maintaining fibers*) have been investigated. However, no transmission systems using such fibers have been realized so far. Ordinary single-mode fibers may be sufficient for present and future applications in subscriber loops. The existence of a cutoff in a fiber is similar to the situation of a metallic hollow waveguide used in a microwave frequency. However, the cutoff value V_C is the cutoff for the next mode of the fundamental mode in a single-mode fiber. A single-mode transmission is realized when $V < V_C$. In metallic hollow waveguides, a single-mode transmission is realized in the frequency band $f_{c1} < f < f_{c2}$, where f_{c1} is the cutoff of the fundamental mode and f_{c2} is the cutoff of the next mode [8]. For example, the fundamental TE_{10} mode for a metallic waveguide with a rectangular cross section of $a \times b$ ($a > b$) is the single mode in the frequency band $c/(2a) < f < c/a$, where c is the velocity of light.

Many modes can propagate when $V > V_C$. V takes a large value when core diameter $2a$ and Δ are large. These fibers are called *multimode fibers*. Multimode fiber has a wider baseband response when α in (2.1) is about 2. Since (2.1) shows a parabolic refractive index profile for a fiber with $\alpha \approx 2$, the fiber is called *graded-index multimode fiber*. The typical values of commonly used graded-index multimode fibers are shown in Figure 2.1. Since the core diameter is 50 μm, it is easy to couple it to laser light and to splice it.

To obtain a low-loss fiber, material loss of a fiber must be decreased. Fiber loss, including material intrinsic loss, $\alpha(\lambda)$, is expressed by (2.5) as a function of wavelength λ:

$$\alpha(\lambda) = \frac{C_1}{\lambda^4} + C_2 + A(\lambda) \tag{2.5}$$

where C_1 and C_2 are constants and $A(\lambda)$ is a function of λ. The first term is due to Rayleigh scattering loss. Microfluctuation of the refractive index, which is smaller than the light wavelength used, causes Rayleigh scattering. The origin of microfluctuation is the fluctuation of density or composition in glass. This scattering is proportional to λ^{-4} and fiber loss decreases greatly for longer wavelengths when this scattering is dominant. Constant C_1 equals 0.6 to 0.8 (μm^4 dB/km) for today's single-mode fibers. The second term is the loss due to imperfection in fiber structure, such as microbending and core-cladding imperfection. This loss mechanism is wavelength-independent. Constant C_2 is very small for today's fibers because of fiber manufacturing technology improvement. $A(\lambda)$ represents the loss of impurities and intrinsic *ultraviolet* (UV) and *infrared* (IR) absorptions. OH and metal ions are the major impurities for silica fibers. Metal ions are not a problem for today's fibers due to fiber manufacturing technology improvement. Loss peaks due to OH exist at wavelengths of 0.95, 1.13, 1.24, 1.39, 1.90, and 2.22 μm in the 0.9- to 2.5-μm wavelength band [9]. OH is also eliminated for good fibers, but a slight amount still exists at present. OH is an important factor for the lightwave system designer, even when good fibers are being used. OH penetration into a fiber causes fiber-loss increase.

It affects long-term reliability on fiber loss. It is known that OH is also made by diffused hydrogen gas and constituent oxygen atoms after optical cable installation. Loss values due to this mechanism depend on the dopant type and optical cable materials. Today's optical transmission systems are constructed with the consideration of this mechanism, and they work well. UV absorption originates from the electronic band gap transitions, and the loss due to this absorption decreases with longer wavelengths. IR absorption, which originates from multiphonon absorption (molecular vibration), increases with longer wavelengths. The loss minimum wavelength depends on the first and third terms in (2.5). Typical transmission losses as a function of single-mode fibers are shown in Figure 2.2. Two curves represent OH existing and OH free single-mode fibers, respectively. The reported minimum transmission loss is 0.15 dB/km at 1.55 μm [10].

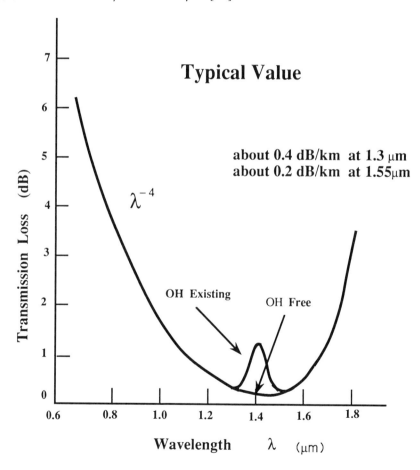

Figure 2.2 Spectral loss of optical fiber.

Fiber structural design is also important for obtaining low-loss fiber and low-loss optical cable. Light injected from a light source, such as a laser diode, is guided by a core of a fiber. Most guided light exists in a core, but some of the light is in a cladding [5]. Therefore, material loss of cladding, especially near a core, as well as that of a core, must be low to obtain low-loss fibers. It is also important that the cladding diameter be suitably designed [11]. Since some small part of light exists in a cladding, the coupling of light between cores occurs when the distance between cores is small. This is the principle of fused-fiber couplers, and it is analogous to a microwave directional coupler using strip lines.

There are two important parameters to consider when designing optical transmission systems using optical fibers: transmission loss and dispersion. For high-bit-rate transmission, dispersion must be taken into consideration. Dispersion is related to group delay τ (propagation time of light through a fiber). The group delay τ per unit length is defined by the following equation.

$$\tau = \frac{1}{V_g} = \frac{d\beta}{d\omega}$$

(2.6)

where V_g is the group velocity and β and ω are phase constant and angular frequency, respectively. Since many modes can propagate and the velocity of each mode differs, modal dispersion takes place in a multimode fiber. β is a function of the refractive index of the fiber material. The refractive index of core and cladding materials is wavelength-dependent. This dispersion is called *material dispersion*. β is also affected by a fiber guide structure, and it is wavelength-dependent. The dispersion due to the waveguide structure is called *waveguide dispersion*. In multimode fibers, modal dispersion is dominant. For ordinary usage of single-mode fibers, only one mode exists. In this case, there is no modal dispersion and both material and waveguide dispersions are important in single-mode fibers.

First, we show the calculated results for a multimode fiber with an index profile expressed by (2.1). The results are

$$\tau(m) = \frac{N_0}{c} \left[1 + \Delta \left(\frac{\alpha - 2 - y}{\alpha + 2} \right) \left(\frac{m}{M} \right)^{2\alpha/\alpha + 2} \right.$$
$$\left. + \Delta^2 \left(\frac{3\alpha - 2 - 2y}{2\alpha + 4} \right) \left(\frac{m}{M} \right)^{4\alpha/\alpha + 2} \right]$$

(2.7)

$$N_0 = n_0 - \lambda \frac{dn_0}{d\lambda}$$

(2.8)

$$y = -\frac{2n_0}{N_0} \frac{\lambda}{\Delta} \frac{d\Delta}{d\lambda}$$

(2.9)

where m is the mode number and M is the maximum mode number. These equations were derived using the Wentzel-Kramers-Brillouin (WKB) method [12]. It is known from these equations that the group delay difference among modes can be decreased when $\alpha = 2 + y$. For multimode fibers with $\alpha = 2 + y$, the group delay difference is proportional to Δ^2 and results in a small value (this means fibers with wide bandwidth). Parameter y is related to the wavelength dependence of the refractive index. Therefore, the optimum α value, $\alpha_{opt} = 2 + y$, is dependent on wavelength, fiber parameters, and material. For a 1.3-μm wavelength, $\Delta = 1\%$, and for GeO_2 dopant fiber, $\alpha_{opt} = 1.9$. The physical image of a wide bandwidth of multimode fibers with $\alpha \approx 2$ (graded-index multimode fibers) is as follows. We divide fiber modes into two mode groups: higher modes and lower modes. Lower modes propagate near the fiber axis, where refractive index is high. Because of the high refractive index, propagation speed is slow compared to that of higher modes. However, propagation distance for lower modes is short because higher modes have a large propagation angle and they propagate from an area near the axis to an area near the cladding. With slow speed and short distance for lower modes and high speed and long distance for higher modes, nearly the same delay is realized.

Next, the dispersion of single-mode fibers will be considered. First, the impulse light propagates a unit-length fiber. Then the impulse light broadens to light with width $\delta\tau$:

$$\delta\tau = (\sigma_m + \sigma_w)\delta\lambda \qquad (2.10)$$

where σ_m and σ_w are material and waveguide dispersions, respectively. $\delta\lambda$ is the spectral width of the signal light. Although both material and waveguide dispersions are important, the value of material dispersion is larger than that of waveguide dispersion in ordinary single-mode fibers. Therefore, material dispersion mainly determines the zero dispersion wavelength λ_0. For ordinary SI fibers, the 1.3-μm wavelength is the zero dispersion wavelength λ_0. Waveguide dispersion is small but it can move λ_0 slightly. Fibers with $\lambda_0 = 1.55$ μm are called *dispersion-sifted fibers* (DSF). Several refractive index profiles, which are different from step-index profiles, have been proposed. A typical dispersion of single-mode fibers is shown in Figure 2.3. The dispersion shown in Figure 2.3 includes both material and waveguide dispersions. To design optical systems, possible transmission bit rates must be evaluated from the dispersion value. It is well known that the frequency response $H(f)$ is the Fourier transform of impulse response $h(t)$.

$$H(f) = \int_{-\infty}^{\infty} h(t)e^{-j2\pi ft}dt \qquad (2.11)$$

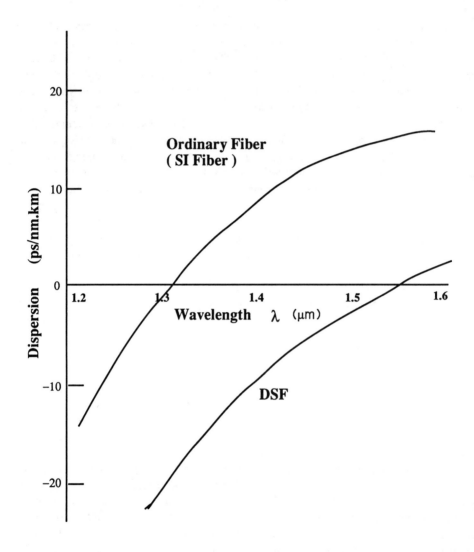

Figure 2.3 Dispersion of single-mode fibers.

In this case, $h(t)$ is replaced by a light pulse with width $\delta\tau$. To consider the simple case of a small $\delta\tau$ and symmetric $h(t)$, a Taylor's series expansion is applied to (2.11). Then we obtain

$$H(f) \approx \int_{-\infty}^{\infty} h(t)\, dt - 2\pi^2 f^2 \int_{-\infty}^{\infty} t^2 h(t)\, dt = H(0)[1 - 2\pi^2 f^2 \sigma_t^2] \quad (2.12)$$

σ_t is the mean square width of $h(t)$. The 3-dB bandwidth f_{3dB} is obtained from (2.12) when $\pi^2 f^2 \sigma_t^2 = 1/4$. This results in

$$f_{3dB} = 1/(2\pi \sigma_t) \quad (2.13)$$

Roughly speaking, the bit rate B is nearly equal to f_{3dB}. Since $\delta\tau$ is for a unit length, the bit rate B is roughly estimated by the following equation for fibers with length L. Here, $h(t)$ is assumed to be a rectangular shape.

$$BL \approx \frac{0.55}{(\sigma_m + \sigma_w)\,\delta\lambda} \quad (2.14)$$

For example, if a 10-km fiber with $\sigma_m + \sigma_w = 10$ ps/km·nm and a light source with $\delta\lambda = 5$ nm, then $B \approx 1.1$ Gb/s. With a narrower light source, $\delta\lambda = 0.05$ nm, $B \approx 110$ Gb/s.

2.1.2 Optical Cable

Optical fibers are used in a form of optical cable for optical transmission systems. There are many proposed cable structures being investigated. Roughly speaking, there are two types of cables used today from the viewpoint of coated optical-fiber structures: one is assembled of mono-coated fiber and the other is of fiber ribbon. The structure of coated optical fiber is shown in Figure 2.4. In mono-coated fiber, a fiber is individually coated with plastics. A bare fiber with 125 μm (= 0.125 mm) is primary-coated and secondary-coated. The coated diameter typically ranges from 0.5 to 1.0 mm. The mechanical strength of mono-coated fiber in the axial direction is around 8 kg. Fiber ribbon has the structure of aligned fibers in a row, and the fibers are secondary-coated to form a ribbon shape. Figure 2.4 shows the case of 8-fiber ribbon. At present, 4-fiber ribbon, 8-fiber ribbon, and 12-fiber ribbon are used. Optical-fiber cable is shown schematically in Figure 2.5. Optical-fiber cable (abbreviated as *optical cable*) with high-count fiber is composed of units. A unit contains several mono-coated fibers or fiber ribbons. The outlook of optical cables is similar to that of metallic cables. Similar treatment is possible for optical cables in the case of cable installation.

Figure 2.6 shows the flow of the optical transmission system construction. The overall system design is first. The system parameters are number of subscribers,

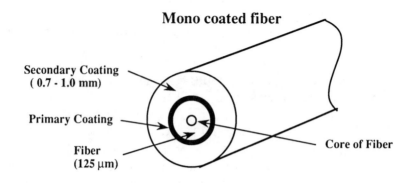

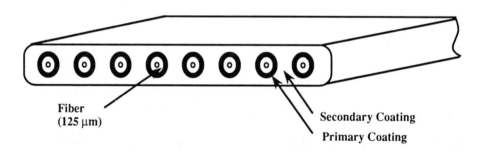

Figure 2.4 Structure of coated optical fiber.

wavelength, bit rate, transmission length, light source device, optical detector device, and so on. Next is the cable route design, considering the scattering pattern of subscribers, the state of the outside plant, the cable piece length, cable access points (these points are used to connect TO with subscribers), and cable installation technique. Then cable network construction is actually started. Cable installation, fiber splicing using splicing techniques or optical connector techniques, and cable jointing using a cable closure or a cable box is done sequentially. To construct the required cable network, this sequence is repeated. After the cable route is constructed, the

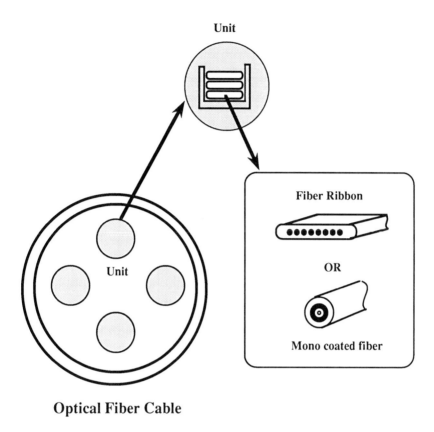

Figure 2.5 Structure of optical-fiber cable.

route is tested by measuring the optical loss of the route. An *optical time domain reflectometer* (OTDR) is commonly used. In OTDR, a short optical pulse is injected into spliced fibers. The injected light is back-scattered by the Rayleigh scattering mechanism. We can evaluate fiber loss and splice loss (connector loss) from one cable route end (such as a telephone office). Sometimes the baseband bandwidth of the route is also measured. For modest-bit-rate systems, this measurement is skipped when using a single-mode fiber. Along with the cable network, transmission equipment is installed. Finally, an overall optical transmission system test for the route is executed.

Although the outlook of optical cables is similar to that of metallic cables, optical cables are superior to metallic cables. A comparison of optical and metallic cables is shown in Table 2.1. The typical outer diameter and weight for 100-fiber and 1000-fiber cables are shown in this table. The outer diameter is from 1/3 to

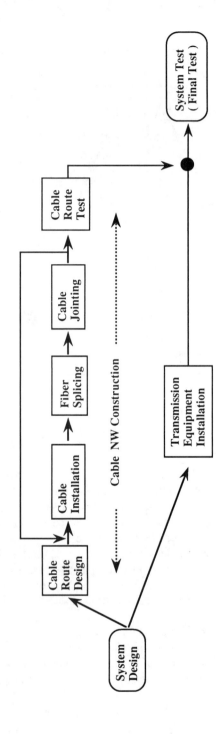

Figure 2.6 System construction flow.

Table 2.1
Comparison of Optical-Fiber Cable and Metallic Cable

	100-Fiber Optic Cable	1,000-Fiber Optic Cable	Optic versus Metallic Cable*
Outer diameter	~ 15 mm	~ 40 mm	~ 1/3–1/2
Weight	~ 0.2 kg/m	~ 1.4 kg/m	~ 1/15–1/8
Transmission loss†			~ 1/1,000

* When compared to a 100-pair or 1,000-pair metallic cable.
† Transmission bit rate 100 MB/s is assumed.

1/2 and the weight is from 1/15 to 1/8 of the outer diameter and weight of metallic cables with the same number of copper pairs. The transmission loss of an optical cable is only about 0.1% of that of a metallic cable at the bit rate of 100 Mb/s.

2.2 FIBER SPLICE AND CONNECTOR

To make use of the superior character of the low-loss optical fiber in transmission systems, it is very important to realize a low-loss splice or low-loss connector. Typical fiber loss is 0.2 to 0.4 dB/km in long wavelengths. A 0.3-dB splice loss corresponds to about a 1-km fiber loss. As mentioned in Figure 2.1, the fiber core diameter is very small. Submicron accuracy is required to obtain a low-loss splice or connector in the case of single-mode fibers. For ordinary single-mode fibers, splice loss or connector loss L due to lateral displacement d is calculated by the following equation:

$$L = 4.34 \left(\frac{d}{\omega}\right)^2 \quad (\text{dB}) \quad (2.15)$$

where ω is the mode field radius. The equation is derived using a Gaussian approximation of mode field in a single-mode fiber [13]. For lateral displacement $d = 1$ μm, L is calculated to be 0.27 dB for $\omega = 4$ μm. To suppress the loss below 0.25 dB, lateral displacement must be under 1 μm (submicron). Practical loss factors are not only lateral displacement, but also angular displacement and fiber facet quality. The calculated requirement used only for lateral displacement may be slightly indulgent. This accuracy is required outside the plant (a very hard environment when compared to a room or telephone office) and is also required for the long term.

Reflection as well as loss is also a very important factor for some applications, such as high-bit-rate transmission, analog transmission, and coherent transmission systems. Reflections in these systems degrade the system performance. The optical isolator can block the reflection, just like an isolator of microwave systems. However, an optical isolator is expensive at present, and, from an economical viewpoint, it is not desirable to use it in subscriber loop systems. The reflection of splice or connector due to the refractive index difference is roughly estimated by the following well-known equation:

$$R = \left(\frac{n_1 - n_2}{n_1 + n_2}\right)^2 \quad (2.16)$$

This equation is valid for a plain wave, but rough estimation can be possible with this equation.

A comparison of spices and connectors is shown in Table 2.2. A splice is the connection of fibers that will not be frequently reconnected. Examples are fusion and mechanical splices. An optical connector is possible to reconnect easily. Generally speaking, splice loss is smaller than optical connector loss. Reflection due to a splice or connector occurs because of the refractive index difference between the fiber and the spliced part (or connected part). As will be explained later, a fusion splice has very low reflection, and mechanical splices and connectors do not have such low reflection. The values in parentheses in Table 2.2 are typical values for ordinary single-mode fibers.

Table 2.2
Comparison of Splices and Connectors

	Splice	Optical Connector
Examples	Fusion splice Mechanical splice	Single-fiber connector Multifiber connector
Reconnection	Not possible or difficult	Easy
Connection loss	Very low (<0.1 dB)*	Low (0.1 to 0.3 dB)*
Reflection due to connection	Fusion splice has very low reflection (< −70 dB)* Mechanical splice and connectors have similar values (−15 dB to −45 dB)*	

* Values in parentheses are typical values for single-mode fibers.

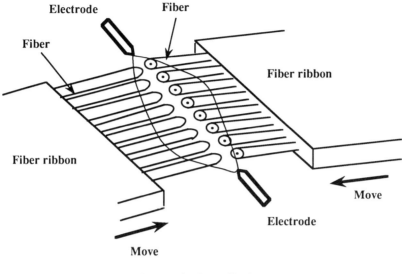

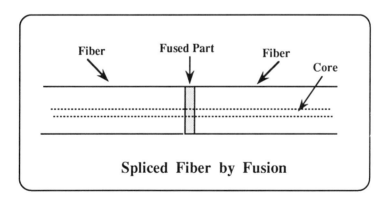

Figure 2.7 Fusion splice of optical fiber.

As examples of splices and connectors, the mass-fusion splice and array connectors are shown in Figures 2.7 and 2.8. In Figure 2.7, the mass-fusion splice and spliced fibers by fusion are shown. The mass-fusion splice is applied to fiber ribbons, and fibers in fiber ribbons are fusion-spliced simultaneously with a single discharge [14]. Splice loss by a mass-fusion splicing machine is obtained under 0.1 dB in the

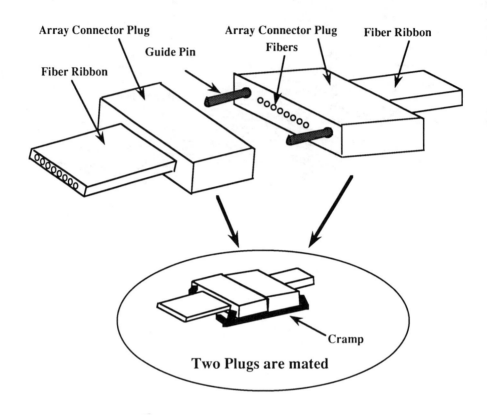
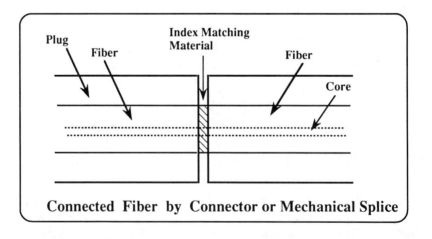

Figure 2.8 Array connector.

field environment. Spliced fibers by fusion are shown in the same figure. The refractive index of the spliced part is slightly different from that of the ordinary fiber part, and it results in the reflection of light. The reflection of a fusion splice is calculated by (2.16) using Δ, such as

$$R \approx \left(\frac{n_1 - n_2}{2n_1}\right)^2 = \frac{\Delta^2}{4} \tag{2.17}$$

In this equation, it is assumed that the refractive index of the spliced part is that of cladding. Therefore, the reflection calculated by (2.17) is overestimated for an ordinary fusion splice. For $\Delta = 1\%$ and $\Delta = 0.2\%$, R is calculated as -46 dB and -60 dB, respectively. Actually, reflection is very small. The measured reflection of fusion-spliced graded-index fiber was reported in [15]. The reflection of twelve cascaded splices with the 5-dB total loss was measured to be about -50 dB, and the reflection of fusion-spliced graded-index fiber with 0.1-dB loss was estimated to be -68 dB. As for a single-mode fiber with a splice loss under 0.1 dB, the reflection is estimated to be well under -70 dB.

Figure 2.8 shows an example of an array connector [16]. This connector is made by using plastic-molding technology. Two plugs are precisely aligned by two guide pins. About 0.3 dB of connection loss on average for 8-fiber ribbons with ordinary single-mode fibers is obtained using index matching material. In the same figure, connected fibers by connectors or by mechanical splice are shown. Usually, index matching material is applied to lower the loss and reflection. The reflection of these connections can be estimated with (2.16). The air gap connector, which uses no index matching material and whose gap is not small, has a reflection with a value of -15 dB. The value is calculated to be about -15 dB, assuming $n_1 = 1$ and $n_2 = 1.46$. With index matching material, no reflection is anticipated by equation (2.16). However, it is difficult to realize ideal index matching because of the fact that the refractive index in a matching material is temperature-dependent and the fact that fiber facet conditions are not ideal. Practically, a reflection of -45 dB is obtained.

Figure 2.9 shows low-reflection techniques for optical connectors without index matching material. The contact state in which two plugs are mated without a gap is called *physical contact*. This is realized by the sophisticated polishing technique, and it realizes both low loss and low reflection because of there being no index difference between two fibers. Practically, a reflection of about -45 dB is obtained because of nonideal fiber facet conditions. Another technology is an obliquely polished connector, which is applied to achieve low reflection. An obliquely polished connector generally has larger loss than ordinary connectors. A reflection of about -60 dB is realized. Connectors mostly used at present have reflection values ranging from -15 to -30 dB.

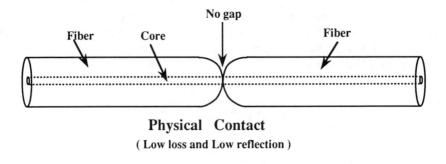

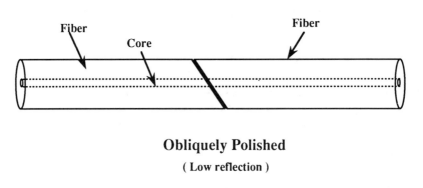

Figure 2.9 Low-reflection and low-loss techniques for optical connectors.

Several fiber splices and connectors are used in optical subscriber loops. Figure 2.10 shows one example. It is assumed that ribbon fiber cables are used for feeder cables. The candidate connection technologies are shown in this figure. These are selected by considering loss, reflection, need for reconnection, connection time, and cost. Generally speaking, the ratio of the number of optical connectors to the number of splices in the subscriber loop will be larger than that in the trunk line because demand for flexibility in the subscriber cable network is higher than that in the trunk line.

2.3 LASER DIODE

The LD is a semiconductor used for a light source. As a result of being a semiconductor, it is small and easy to handle. The LD has been developed along with optical fiber. First, a laser diode for the 0.8- to 0.9-μm wavelength (short wavelength) range was developed, and then one for the 1.3- to 1.6-μm wavelength (long wavelength)

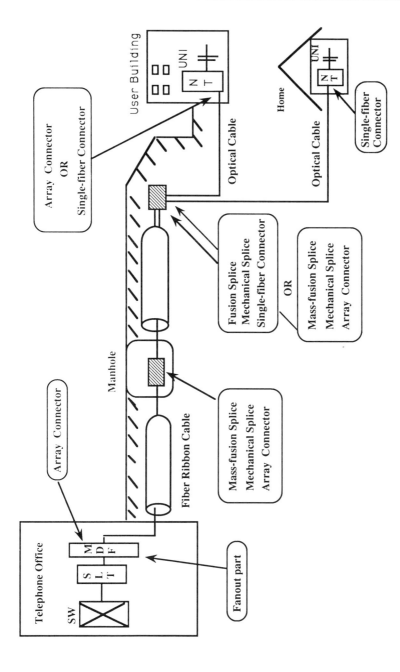

Figure 2.10 Several optical-fiber connection techniques in optical subscriber loops.

range. It is more suitable than the LED for systems using single-mode fibers because of its higher coupling efficiency (coupling between a laser and a single-mode fiber). When constructing an optical transmission system, the following laser parameters must be taken into consideration: wavelength, lasing mode number (single-mode lasing or multimode lasing), modulation characteristics, and linewidth. These parameters are closely related to optical transmission system parameters and are discussed in Section 2.6.

There are many LDs developed for transmission systems. For the 0.8- to 0.9-μm wavelength (short wavelength) range, LDs with AlGaAs for the active layer and GaAs for the substrate material have been developed. For the 1.3- to 1.6-μm wavelength (long wavelength) range, LDs with InGaAsP for the active layer and InP for the substrate material have been developed. Ordinary LDs are buried double-heterostructure, which means that an active layer is surrounded by cladding materials and they form a heterojunction. Through the heterojunction, carrier confinement and optical mode confinement in an active layer are realized. The cladding materials have a larger band gap and the larger band gap materials have a lower refractive index. Carrier confinement is the result of this band gap difference. Optical mode confinement is the result of forming a guiding wave structure with a high refractive index for the active layer and a low refractive index for the cladding layer. Figures 2.11 to 2.13 show the fundamental structures and lasing conditions for fundamental LDs such as the *Fabry Perot–laser diode* (FP-LD), the distributed feedback–laser diode DFB-LD, and the *distributed Bragg reflector*–laser diode (DBR-LD).

2.3.1 FP-LD

The FP-LD has a Fabry-Perot resonator formed by two facets. Reflections of light generating in an active layer take place at facets due to the refractive index difference between laser material and air. Multiple traveling in an active layer caused by reflections at facets results in lasing. The gain after traveling one round trip in the resonator (cavity) is given by

$$\exp(2\Gamma gL) \qquad (2.18)$$

And the loss is

$$\frac{1}{R^2}\exp[2\alpha_1\Gamma L + 2\alpha_2(1-\Gamma)L] \qquad (2.19)$$

where g, Γ, L, α_1, α_2, and R are the optical gain for unit length, the mode confinement factor, the laser length, the loss for active layer, the loss for cladding layer,

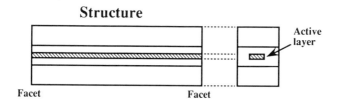
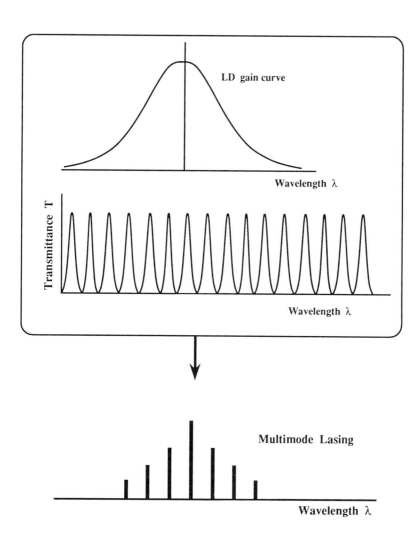

Figure 2.11 FP laser diode.

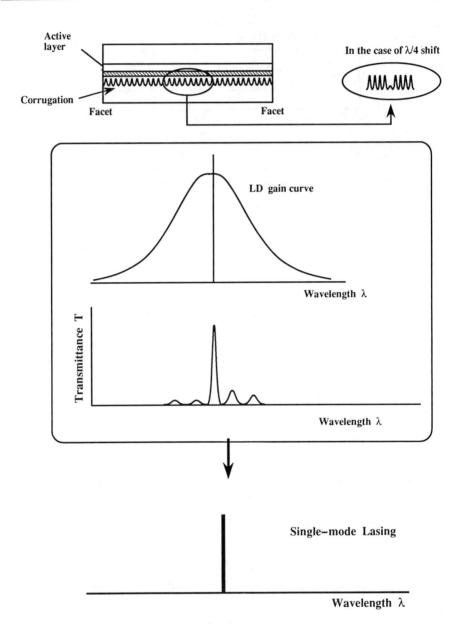

Figure 2.12 DFB laser diode.

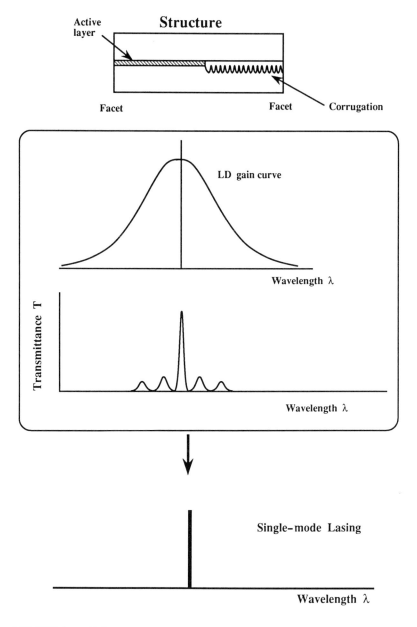

Figure 2.13 DBR laser diode.

and the facet reflectivity, respectively. Γ is the ratio of optical field inside the active layer to total optical field. We assume that $\alpha_1 = \alpha_2 = \alpha$ for simplicity, and then (2.19) is

$$\frac{1}{R^2}\exp(2\alpha L) \qquad (2.20)$$

Threshold gain g_{th} for lasing is obtained by equating (2.18) with (2.20):

$$\Gamma g_{th} = \alpha + \frac{1}{L}\ln\left(\frac{1}{R}\right) \qquad (2.21)$$

Gain g is given by

$$g = A_g(N_e - N_0) \qquad (2.22)$$

where A_g, N_e, and N_0 are the differential gain coefficient ($= dg/dN_e$), the injected carrier density, and the carrier density where $g = 0$, respectively. The threshold current density is obtained from (2.21) and (2.22) as

$$J_{th} = \frac{e(2d)N_{th}}{\tau} = \frac{e(2d)}{\tau}\left[N_0 + \frac{\alpha + \frac{1}{L}\ln\left(\frac{1}{R}\right)}{\Gamma A_g}\right] \qquad (2.23)$$

where e, $2d$, N_{th}, and τ are the electron charge, the active layer thickness, the injected carrier density at threshold, and the carrier life time, respectively. To lower the threshold current J_{th}, the active layer thickness $2d$ must be thin. It looks like J_{th} is the minimum when $2d = 0$. This simple result is wrong. Equation (2.23) contains the mode confinement factor Γ. For a thinner active layer, Γ becomes a small value, which makes J_{th} large. There are optimum values for $2d$. Typical values for $2d$ are around 0.1 to 0.2 μm. For the Fabry-Perot resonator, the transmittance of the resonator has a property as shown in Figure 2.11. Since a semiconductor laser has a broad gain curve as shown in the same figure, lasing occurs at the wavelength with the condition that the round-trip phase is in-phase. Since the time dependence is $\exp(j\omega t)$, the in-phase condition for the round trip is

$$\omega\frac{2L}{V_g} = 2\pi m \qquad (2.24)$$

where V_g, ω, and m are the group velocity, the angular frequency, and the integer (mode number), respectively. Equation (2.24) can be rewritten as

$$\lambda = \frac{2n_g L}{m} \qquad (2.25)$$

where n_g is the group index of an active layer. Lasing occurs at several discrete wavelengths, which results in multimode lasing.

2.3.2 DFB and DBR LDs

The DFB-LD has corrugation along the active layer. DBR-LD has corrugation that is not along the active layer. Although the operation of the DFB-LD is complex, the fundamental operation of the DFB-LD and DBR-LD are understood by Bragg reflection. The built-in corrugation causes the Bragg reflection, and the Bragg wavelength λ_B is given by [17]

$$\lambda_B = \frac{2n_e \Lambda}{p} \qquad (2.26)$$

where Λ, n_e, and p are the period of the grating, the effective refractive index of the mode, and the integer order of the grating, respectively. The mode at the Bragg wavelength λ_B has the lowest loss. Because of (2.21), the threshold gain is the lowest for modes with the lowest loss. Therefore, the mode lasing at λ_B is dominant, leading to single-mode lasing. This is shown in the middle of Figures 2.12 and 2.13. The dispersion given by (2.10) is proportional to the spectral width of signal light. Multimode lasing broadens the spectral width. Therefore, single-mode lasing is important for high-bit-rate transmission. Sometimes, multimode lasing occurs under high-speed modulation for a laser which has a single-mode lasing character under the condition of no modulation or low-speed modulation. A stable single-mode lasing laser is called a *dynamic single-mode* (DSM) laser [18]. DSM lasing is achieved by introducing a $\lambda/4$ phase shift into the grating structure of a DFB laser (shown in Figure 2.12), while a DFB laser without the shift has inherently a two-mode lasing property [19].

2.3.3 Modulation Characteristics

The modulation characteristics of LDs by direct modulation schemes determine the upper modulation speed (modulation frequency). It is also possible to modulate light from LDs by using external modulators, such as a LiNbO$_3$ modulator. However, this approach is complex and expensive for subscriber transmission systems when compared to the direct modulation schemes. Modulation responses of the LD are shown in Figure 2.14. The upper part of this figure shows intensity response, and this is used for *intensity modulation* (IM). The bottom part shows frequency deviation, and this is used for *frequency shift keying* (FSK) modulation in coherent transmission. When a small signal modulation current is applied to a laser diode biased at a given point, the output light is modulated by a signal. For digital transmission, the bias point is below the threshold, or zero (no biasing). For analog transmission, the bias point is above the threshold because no deformation of signal is desirable. It is well known that the frequency response has a peak at the relaxation frequency f_0, as shown in the upper part of Figure 2.14 [20,21]. By solving the following rate equations (for photons and electrons), the relaxation frequency f_0 is obtained.

The rate equations are

$$\frac{dN}{dt} = \frac{I}{eV} - \frac{N}{\tau_c} - A_g V_g (N - N_0) S \tag{2.27}$$

$$\frac{dS}{dt} = A_g V_g \Gamma (N - N_0) S - \frac{S}{\tau_p} + \beta \Gamma \frac{N}{\tau_c} \tag{2.28}$$

where N, S, I, e, V, A_g, N_0, V_g, and Γ are the carrier density inside the cavity, the photon density inside the cavity, the injected current, the electron charge, the volume of active layer, the differential gain coefficient, the carrier density at transparent condition (the same as N_0 in (2.22)), the group velocity inside the cavity, and the mode confinement factor, respectively. β is the spontaneous emission factor, which is the fraction of spontaneous emission entering the lasing mode. τ_c and τ_p are the carrier lifetime and the photon lifetime, respectively. Photon lifetime τ_p is determined by both the facet (mirror) loss and the active layer loss; that is,

$$\tau_p = V_g^{-1} \left[\alpha - \frac{1}{L} \ln(R) \right]^{-1} \tag{2.29}$$

where V_g is the group velocity and other symbols are the same as those in (2.21).

The obtained relaxation frequency f_0 is expressed approximately by [22–24]:

$$f_0 = \frac{1}{2\pi} \sqrt{\frac{A_g V_g S}{\tau_p}} \tag{2.30}$$

The value of f_0 for high-speed LDs was measured to be 2 to 7 GHz for the bias current $I_b = 15$ to 25 mA [22]. As indicated by equation (2.30), f_0 is proportional

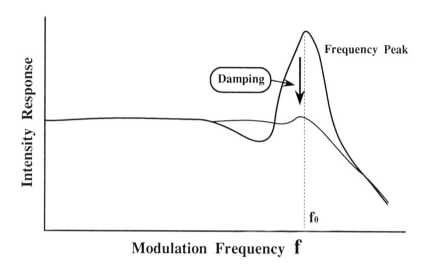

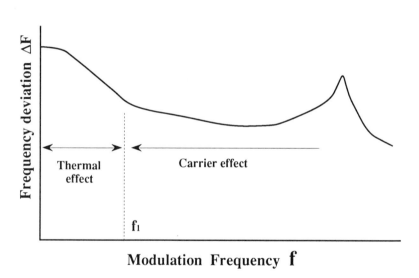

Figure 2.14 Modulation response of LD.

to the square root of S. This means that the relaxation frequency f_0 increases along with the optical power or the bias current. The value of f_0 for ordinary, commercially available 1.3-μm wavelength Fabrey-Perot laser diodes (InGaAsP) is about 2 GHz. The maximum modulation frequency for intensity modulation is determined to be [24]

$$f_{\max} = 2\sqrt{2}\,\pi f_0^2/\gamma \qquad (2.31)$$

where γ is the damping factor. One example value of the reported $f_{\max}$ for high-speed LDs is about 15 GHz [24].

The bottom of the same figure is the frequency deviation as a function of modulation frequency (FM response of LD). In an ideal FSK, frequency only deviates without any variation of light intensity. The frequency deviation ΔF was caused by two effects: thermal and carrier [25]. The thermal effect is the change of refractive index in an active layer due to the temperature, dn/dT, where the symbol n is the refractive index in an active layer. The carrier effect is the refractive index change caused by the carrier, dn/dN_e, where N_e is the injected carrier. Here, we define the frequency f_1 as the boundary frequency of these two effects, which is indicated in Figure 2.14. The reported f_1 frequency is about 10 MHz and the reported normalized frequency deviation $\Delta F/\Delta I$ is about 3 GHz/mA for frequencies well below f_1, and 0.1 GHz/mA for frequencies above f_1 (in the region of the flat frequency response) for AlGaAs LDs [25]. Those values of recent multielectrode InGaAsP DFB and DBR LDs were also measured [26,27]. The reported $\Delta F/\Delta I$ is 1 to 2 GHz/mA for a multielectrode DFB-LD [26]. A flat FM response with a value of 0.3 GHz/mA over a 1-GHz bandwidth was reported in [27] for a multielectrode DBR-LD.

2.3.4 Linewidth of LD

The linewidth of a laser is a very important parameter when considering optical frequency-division multiplexing systems. The origin of the width is spontaneous emission (a quantum mechanical phenomenon) and was calculated by Schawlow and Townes [28]. For a laser diode, the measured width is broader than that calculated by Schawlow and Townes. The change of an imaginary part of the refractive index, which is related to gain g, is considered to be caused by the change of the injected carrier density N_e. Then it, in turn, changes the real part of refractive index n. The linewidth was modified by a factor of $(1 + \alpha^2)$, where α is the linewidth enhancement factor [29]. The modified linewidth δf is expressed by

$$\delta f = \frac{V_g^2\, hf\, n_{\mathrm{sp}}}{8\pi P}\, \alpha_m\,(\alpha_m + \alpha_L)(1 + \alpha^2) \qquad (2.32)$$

$$\alpha = -\frac{4\pi}{\lambda}\frac{(dn/dN_e)}{(dg/dN_e)} \qquad (2.33)$$

where v_g, n_{sp}, α_m, and α_L are the group velocity, the inversion parameter (or the spontaneous emission factor), the facet loss (or the mirror loss), and the waveguide loss, respectively. α_L is the same as α in (2.20). α_m is expressed by

$$\alpha_m = \frac{1}{L} \ln\left(\frac{1}{R}\right) \tag{2.34}$$

Generally, (2.32) indicates that δf decreases as an optical power increases. The value of the linewidth enhancement factor α is about 2 to 6 and the broadening factor $(1 + \alpha^2)$ is about 5 to 37. The reported linewidth δf is from several megahertz to several gigahertz for a laser diode without external devices. Using a grating attached externally to a laser diode to form the extended cavity, the value of δf can be decreased under 100 kHz.

2.3.5 Chirping

The frequency chirping is the lasing frequency change due to the variation of optical power. This characteristics is important for high-bit-rate transmission because the combination of this chirping and fiber dispersion cause the degradation of short-pulse transmission.

2.3.6 Equivalent Circuit

Figure 2.15 shows the equivalent circuit of a laser diode [30,31]. The device is a kind of diode with a forward bias when used for a light source. The capacitances C and C_s are the intrinsic junction capacitance and stray capacitance due to external circuit and LD package, respectively. Inductance L comes from a lead wire used for the electrode.

Sample values for these circuit parameters are $r = 5\Omega$, $C = 20$ pF, $L_{lead} = 1$ nH, $C_s = 1$ pF. Since r is small, external resistance is required for impedance matching when driving it through a 50Ω coaxial cable.

2.4 PHOTODIODE

The photodiode is a semiconductor that is used as a light detector. The *avalanche photodiode* (APD) is a kind of photodiode with internal avalanche multiplication. From the multiplication viewpoint, it is like a photomultiplier, which is a tube photodetector. Photodiodes are generally used with reverse voltage. The most important photodiode parameters include responsivity R, junction capacitance C_t, dark current I_d, and x. The parameter x is related to the excess noise factor for APD.

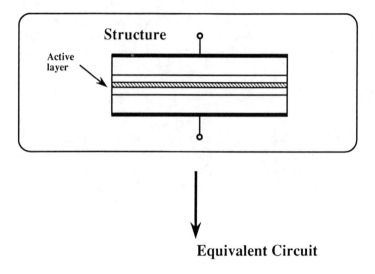
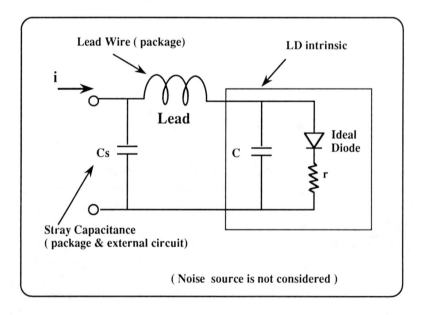

Figure 2.15 Equivalent circuit of LD.

2.4.1 Responsivity and Photocurrent

The absorption of photons in a PD produces electron-hole pairs, resulting in current in external circuits. The current produced by photons is expressed by

$$I_p = \eta \frac{e}{hf} P = RP \qquad (2.35)$$

$$R = \eta \frac{e}{hf} \qquad (2.36)$$

where h, f, e, η, P, and R are Plank's constant (6.63×10^{-34} J·s), the frequency, the electron charge, the quantum efficiency, the optical power incident on the detector, and the responsivity, respectively. The product hf is the photon energy. The responsivity R increases for longer wavelengths when the quantum efficiency is the same. R is expressed simply for the fixed wavelength as follows. $R = 1.05\eta$ for 1.3-μm wavelength. $R = 1.25\eta$ for 1.55-μm wavelength.

For an APD, the current produced by photons is expressed as

$$I = \langle M \rangle \eta \frac{e}{hf} P = \langle M \rangle I_p \quad (APD) \qquad (2.37)$$

where M is the avalanche gain and $\langle M \rangle$ is the mean value of M. The avalanche gain process of an APD is a statistical process. The random feature of the carrier multiplication process introduces noise, which is discussed later. The current i from a PD or APD is composed of I_p (or $\langle M \rangle I_p$) and I_d, such as

$$i = I_p + I_d \quad (PD) \qquad (2.38)$$

$$i = \langle M \rangle I_p + I_d \quad (APD) \qquad (2.39)$$

I_d is the dark current, which flows even when there is no light incident on the detector. The temperature dependence of the dark current is approximately expressed as $\exp(-E_g/2kT)$ or $\exp(-E_g/kT)$, where E_g, k, and T are the energy gap, the Boltzmann constant (1.38×10^{-23} J/k), and the temperature, respectively. The typical values of the dark current I_d in room temperature are 0.05 nA for a Si pin-PD, 5 μA for a Ge pin-PD, and 0.1 nA for an InGaAs pin-PD under the condition of a few volts in reverse direction.

When compared to a pin-PD and APD from the viewpoint of designing an optical receiver, an APD is superior with regard to the receiver sensitivity. However, an APD requires higher voltage for biasing (e.g., about 70V for an InGaAs APD), while a few volts are required for an InGaAs pin-PD.

2.4.2 Wavelength Dependence of *R*

Wavelength dependence of responsivity for three materials such as Si, Ge, and InGaAs, which are commonly used, is shown in Figure 2.16. The data are plotted for commercially available pin-PDs for $V_R = 1V$ (V_R is the bias voltage in the reverse direction). Si is used for the short-wavelength range, and InGaAs for the long-wavelength

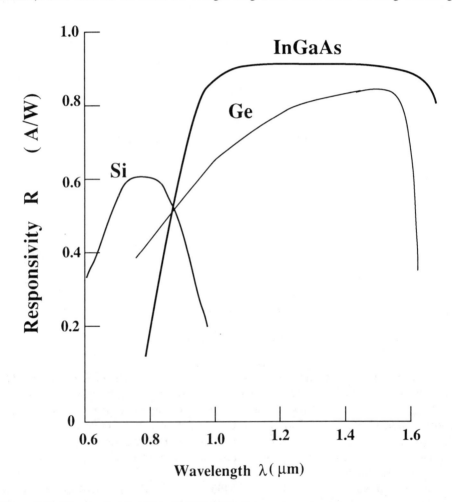

Figure 2.16 Wavelength dependence of responsivity *R*.

range. The responsivity R increases for longer wavelengths for the same quantum efficiency η. Nearly the same values of η, 0.8 to 0.9, are realized for Si and InGaAs.

2.4.3 Equivalent Circuit

The structure and equivalent circuit for pin-PD are shown in Figure 2.17. The capacitance C_t is the junction capacitance, which determines the frequency response characteristics of PDs. The capacitance C_t decreases slightly for larger bias voltage in the reverse direction. The capacitance for parallel plates is expressed as $C = \varepsilon S/d$, where ε, S, and d are the dielectric constant, the area of plate, and distance between two plates, respectively. Analogous to these parallel plates, C_t in Figure 2.17 is thought to have a relationship with a photodetection diameter D of a PD. The straightforward analogy indicates that C_t is proportional to D^2. The capacitances C_t for commercially available PDs and APDs are shown in Figure 2.18 as a function of D. The dotted lines in this figure indicate that C_t is proportional to $D^{1.7}$, which is close to D^2. The typical value of an InGaAs pin-PD or InGaAs APD used for optical transmission with single-mode fibers is about 1 to 2 PF because the required diameter D for this application is not large.

2.4.4 Noise

The photodetection process is a statistical process. The arrival times of photons to a PD are random. It is known that laser light has a coherent state, and the randomness of light with a coherent state follows Poisson statistics [32]. The fluctuation in the current, which is caused by Poisson statistics, is referred to as *shot noise*. When noises of the electric preamplifier and load resistance, which follow a PD or an APD, are not low, the dominant noise in an optical receiver is this thermal noise. On the other hand, the shot noise of a PD or an APD is intrinsic and important for designing a low-noise optical receiver. To obtain the noise of a PD and an APD, we use stochastic mathematics [33]. First, the Poisson distribution $P(n)$ is expressed as

$$P(n) = \frac{(\Lambda T)^n e^{-\Lambda T}}{n!} \tag{2.40}$$

where n, Λ, and T are the photon number, the intensity function, and the time interval $[0,T]$, respectively. The product ΛT is the mean photon number for time interval T. Electrons generated by photons also follow the Poisson distribution $P(n)$ with the mean electron number $\eta \Lambda T$, and it is expressed by

$$P(n) = \frac{(\Lambda \eta T)^n e^{-\Lambda \eta T}}{n!} \tag{2.41}$$

Structure

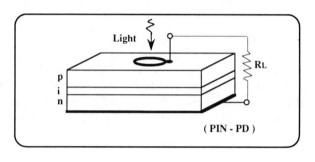

Equivalent Circuit

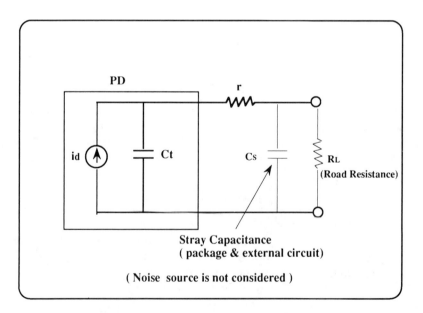

Figure 2.17 Equivalent circuit of PD.

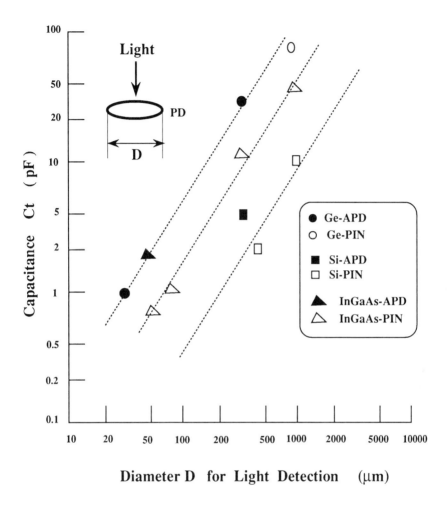

Figure 2.18 Relationship between capacitance and detection diameter.

where η is the quantum efficiency. Second, using the above equation, the mean and the variation of the photocurrent are calculated. The generated photocurrent $i_n(t)$ is

$$i_n(t) = \sum_{j=1}^{n} e\, \delta(t - t_j) \tag{2.42}$$

where e and $\delta(t)$ are the electron charge and the Dirac delta function, respectively. The mean current is

$$\bar{i} = \sum_{n=0}^{\infty} P(n)\, i_n(t) = \eta\, \Lambda\, e \qquad (2.43)$$

The variation, which corresponds to noise, is

$$\langle i^2 \rangle = \langle (i - \bar{i})^2 \rangle = 2\, \eta\, \Lambda\, e^2 \int_0^{\infty} df \qquad (2.44)$$

For the frequency bandwidth limited case, the following mean square noise for a PD is obtained from (2.44) using the equation $I_p = \eta \Lambda e$:

$$\langle i^2 \rangle = 2eI_p\, \Delta f \quad \text{(PD)} \qquad (2.45)$$

where Δf is the frequency bandwidth. This is the well-known result for Poisson statistics. For an APD, the following equation, considering photocurrent multiplication, is derived [34]:

$$\begin{aligned}\langle i^2 \rangle &= 2eI_p \langle M^2 \rangle\, \Delta f \\ &= 2eI_p \langle M \rangle^2 F(M)\, \Delta f \quad \text{(APD)}\end{aligned} \qquad (2.46)$$

where

$$\begin{aligned}F(M) &= \langle M \rangle \left[1 - (1-k)\left(\frac{\langle M \rangle - 1}{\langle M \rangle}\right)^2 \right] \\ &\approx 2(1-k) + k \langle M \rangle \quad \text{(APD)}\end{aligned} \qquad (2.47)$$

$F(M)$ is the excess noise factor due to the random avalanche gain process. The symbol k is the ratio of the ionization coefficients of holes and electrons in the detector junction. Since (2.47) is slightly complex, the following simple equation is often used for a design:

$$F(M) \approx \langle M \rangle^x \quad \text{(APD)} \qquad (2.48)$$

where $x \approx 0.3$ to 0.5 (Si-APD), $x \approx 1$ (Ge-APD), and $x \approx 0.7$ (InGaAs-APD).

The reported x for recent InGaAs-APDs is also 0.7 [35,36]. Using (2.48), equation (2.46) is

$$\langle i^2 \rangle = 2eI_p \langle M \rangle^{2+x} \Delta f \qquad (2.49)$$

As for the dark current, the noises are

$$\langle i_d^2 \rangle = 2eI_d \Delta f \quad \text{(PD)} \qquad (2.50)$$

$$\langle i_d^2 \rangle = 2e(I_m \langle M \rangle^{2+x} + I_n) \Delta f \quad \text{(APD)} \qquad (2.51)$$

where I_m and I_n are constants.

2.5 FIBER COUPLER AND WDM DEVICES

There are many types of couplers and WDM devices. The devices used for single-mode fibers are categorized into three major types: fused fiber, waveguide, and bulk, shown in Figures 2.19 to 2.22. The fused-fiber type is composed of optical fibers that are fused into one by heating and stretching. The waveguide type is made by a silica waveguide [37], glass waveguide [38], and so on. The light guiding structure is formed in a substrate by designing the refractive index, just like a fiber structure. The bulk type uses some extra devices, such as a grating or a prism. The grating on the SiO_2 substrate made by the etching process for WDM is commonly used as a dispersive device. The bulk-type coupler uses a beam splitter such as a half mirror.

2.5.1 Fiber- and Waveguide-Type Devices

The principle of couplers, for both fused-fiber and waveguide types, is the coupling of light. The well-known coupled wave equations for two loss-less waveguides are

$$\frac{da_1}{dz} = -j\beta_1 a_1 + jk_1 a_2 \qquad (2.52)$$

$$\frac{da_2}{dz} = jk_2 a_1 - j\beta_2 a_2 \qquad (2.53)$$

$$a_1(z) = A_1 \exp(-j\beta_1 z) \qquad (2.54)$$

$$a_2(z) = A_2 \exp(-j\beta_2 z) \qquad (2.55)$$

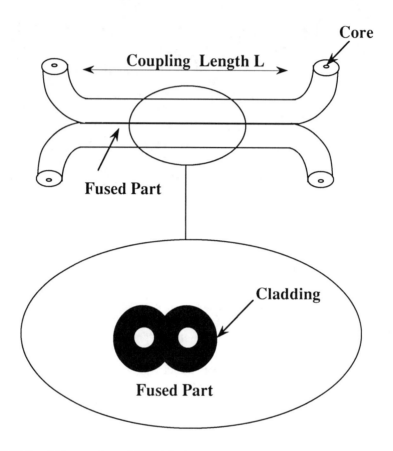

Figure 2.19 Fused-fiber coupler and WDM device.

where a_1 and a_2 are the complex wave amplitudes for waves 1 and 2. A_1 and A_2 are constants. β_1 and β_2 are the phase constants (time dependence of $\exp[j(\omega t - \beta z)]$ is assumed). k_1 and k_2 are the coupling coefficients. Here we treat the case where the two waveguides are indentical. In this case, $\beta_1 = \beta_2 \ (= \beta)$ and $k_1 = k_2 \ (= k)$ are hold. The solution of the above coupled wave equations for $a_2(0) = 0$ are

$$a_1(z) = A \cos(\Delta\beta z) \exp(-j\beta z) \tag{2.56}$$

$$a_2(z) = j A \sin(\Delta\beta z) \exp(-j\beta z) \tag{2.57}$$

$$A = a_1(0) \tag{2.58}$$

$$\Delta\beta = k \tag{2.59}$$

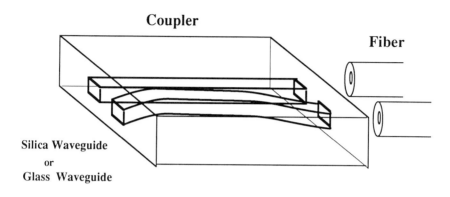

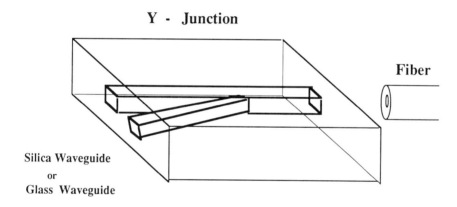

Figure 2.20 Waveguide-type coupler.

$$P_1(z) = |a_1(z)|^2 = A^2 \cos^2(\Delta\beta z) \quad (2.60)$$

$$P_2(z) = |a_2(z)|^2 = A^2 \sin^2(\Delta\beta z) \quad (2.61)$$

When $\Delta\beta z = \pi/2$ (i.e., $kz = \pi/2$), the initial power of waveguide 1 is completely transferred to waveguide 2. It is also known that $P_1(\pi/4k) = P_2(\pi/4k) = 0.5P_1(0)$, which is a 3-dB coupler. The coupling length $L = \pi/4k$ for a 3-dB coupler is applied to both fused-fiber and waveguide types. The coupling coefficient k depends on the waveguide structure, such as distance of two cores, shape of core, and refractive index profile. With an appropriate design of the structural parameters, a desirable k value can be obtained. Nearly the same structures as used for a fused-fiber coupler

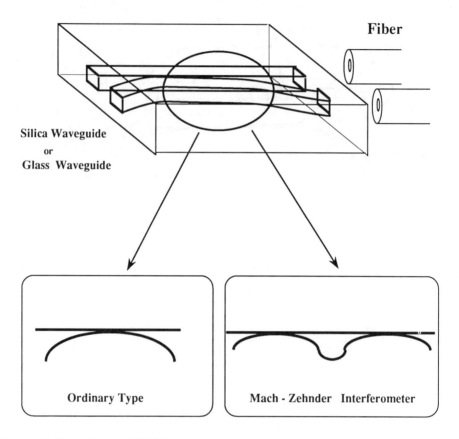

Figure 2.21 Waveguide-type WDM device.

or waveguide coupler are used for WDM devices. The value of the coupling coefficient k for a WDM device is different from that of a 3-dB coupler. The value k is nearly constant for a wavelength range, which is used as a 3-dB coupler. On the other hand, k is variable for a wavelength range in the case of a WDM device. The equation $kL = \pi/2$ holds for wavelength $\lambda 1$, then $P_1(L) = 0$ and $P_2(L) = A^2$. When $kL = \pi$ for wavelength $\lambda 2$, then $P_1(L) = A^2$ and $P_2(L) = 0$. This is the principle of these WDM devices. Some waveguide-type WDM devices use Mach-Zehnder interferometers for easy fabrication because a desirable k value is easily obtained with this interferometer design.

2.5.2 Bulk-Type WDM Device

The bulk-type WDM device uses dispersive devices, which transform the wavelength difference to the light propagation direction difference:

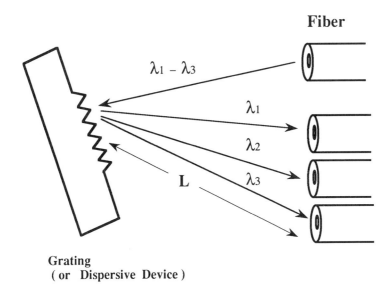

Figure 2.22 WDM device using grating.

$$\Delta x_1 = \left(\frac{d\theta}{d\lambda}\right) \Delta\lambda_1 L \qquad (2.62)$$

$$\Delta x_2 = \left(\frac{d\theta}{d\lambda}\right) \Delta\lambda_2 L \qquad (2.63)$$

and so forth.

$$\Delta\lambda_1 = \lambda_1 - \lambda_2 \qquad (2.64)$$

$$\Delta\lambda_2 = \lambda_2 - \lambda_3 \qquad (2.65)$$

and so forth, where θ is the diffracted angle and L is the length shown in Figure 2.22. Δx_i, $\Delta\lambda_i$ ($i = 1, 2, \ldots$) are the position spaces for fibers and the wavelength space, respectively. This is the principle of the bulk-type wavelength-division multiplexer or demultiplexer.

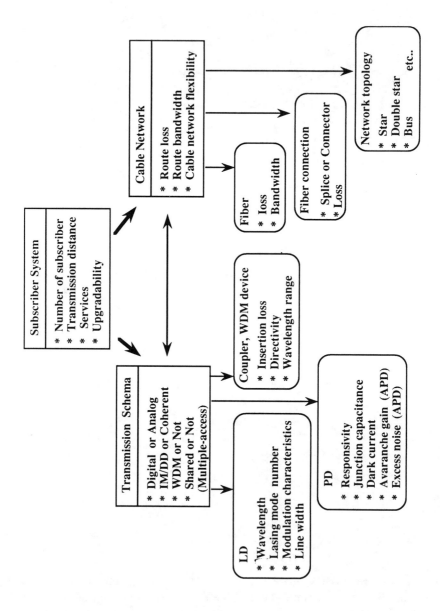

Figure 2.23 Relationship between transmission system and optical device.

2.6 RELATIONSHIP BETWEEN OPTICAL SYSTEMS AND OPTICAL DEVICES

For designing optical subscriber systems, many factors must be considered. The relationship between optical systems and optical devices is shown in Figure 2.23, which is a typical case. For some practical designs, other parameters not shown in this figure must be taken into consideration. Generally speaking, transmission schemes and cable network designs are mutually influenced. Although some factors, such as cost, power feed for subscribers, and easy operation, are not shown in this figure, they are very important for subscriber loops.

REFERENCES

[1] Maiman, T. H., "Stimulated Optical Radiation in Ruby Masers," *Nature*, Vol. 187, 1960, p. 493.
[2] Hayashi, I., M. B. Panish, P. W. Foy, and A. Sumski, "Function Lasers Which Operate Continuously at Room Temperature," *Appl. Phys. Letters*, Vol. 17, 1970, p. 109.
[3] Kao, K. C., and G. A. Hockham, "Dielectric-Fiber Surface Waveguides for Optical Frequencies," *Proc. IEEE*, Vol. 113, 1966, p. 1151.
[4] Kapron, F. P., D. B. Keck, and R. D. Maurer, "Radiation Losses in Glass Optical Waveguides," *Appl. Phys. Lett.*, Vol. 17, 1970, p. 423.
[5] Gloge, D., "Weakly Guiding Fibers," *Appl. Opt. Vol.*, 10, 1971, p. 2252.
[6] Petermann, K., "Constraints for Fundamental-Mode Spot Size for Broadband Dispersion-Compensated Single-Mode Fibers," *Electron. Lett.*, Vol. 19, 1983, p. 712.
[7] Marcuse, D., "Gaussian Approximation of the Fundamental Modes of Graded-Index Fibers," *J. Opt. Soc. Am.*, Vol. 68, 1978, p. 103.
[8] Collin, R. E., "Foundation of Microwave Engineering," McGraw-Hill, 1966.
[9] Kaiser, P., A. R. Tynes, H. W. Astle, A. D. Peoson, W. G. French, R. E. Jaeger, and A. H. Cherin, "Spectral Losses of Unclad Vitreous Silica and Soda-Line-Silicate Fibers" *J. Opt. Soc. Am.*, Vol. 63, 1973, p. 1141.
[10] Yokota, H., H. Kanamori, Y. Ishiguro, G. Tnaka, H. Takada, M. Watanabe, S. Suzuki, K. Yano, M. Hoshikawa, and H. Shimba, "Ultra-Low-Loss Pure-Silica-Core Single-Mode Fiber and Transmission Experiment," *Conf. on Optical Fiber Communication '86 (OFC'86)*, PD3, Atlanta, 1986.
[11] Kashima, N., and N. Uchida," Transmission Characteristics of Graded- Index Optical Fibers With a Lossy Outer Layer," *Appl. Opt.*, Vol. 17, 1978, p. 1199.
[12] Olshansky, R., "Mode Coupling Effects in Graded-Index Optical Fibers," *Appl. Opt.*, Vol. 14, 1975, p. 935.
[13] Marcuse, D., "Loss Analysis of Single-Mode Fiber Splices," *B.T.T.J.*, Vol. 56, 1977, p. 703.
[14] Tachikura, M., and N. Kashima, "Fusion Mass-Splice for Optical Fibers Using High-Frequency Discharge" *IEEE J. of Lightwave Technol.*, Vol. LT-2, 1984, p. 25.
[15] Kashima, N., and I. Sankawa, "Reflection Properties of Splices in Graded-Index Optical Fibers," *Appl. Opt.*, Vol. 22, 1983, p. 3820.
[16] Satake, T., N. Kashima, and M. Oki, "Very Small Single-Mode Ten-Fiber Connector," *IEEE J. of Lightwave Technol.*, Vol. 6, 1988, p. 269.
[17] Yariv, A., "Quantum Electronics," 2nd ed., John Wiley & Sons, 1975.
[18] Suematu, Y., S. Arai, and K. Kishino, "Dynamic Single Mode Semiconductor Lasers With a Distributed Reflector," *IEEE J. of Lightwave Technol.*, Vol. LT-1, 1983, p. 161.

[19] Akiba, S., M. Usami, and K. Utaka, "1.5 μm λ/4-Shifted InGaAsP/InP DFB Lasers," *IEEE J. of Lightwave Technol.*, Vol. LT-5, 1987, p. 1564.

[20] Ikegami, T., and Y. Suematu, "Carrier Lifetime Measurement of a Junction Laser Using Direct Modulation," *IEEE J. Quantum Electron.*, Vol. QE-4, 1968, p. 148.

[21] Ikegami, T., and Y. Suematu, "Resonance-Like Characteristics of the Direct Modulation of a Junction Laser," *Proc. IEEE (Lett.)*, Vol. 55, 1967, p. 122.

[22] Lau, K. Y., N. Bar-Chaim, I. Ury, Ch. Harder, and A. Yariv, "Direct Amplitude Modulation of Short-Cavity GaAs Lasers up to X-Band Frequencies," *Appl. Phys. Lett.*, Vol. 43, 1983, p. 1.

[23] Lau, K. Y., and A. Yariv, "Ultra-High Speed Semiconductor Lasers," *IEEE J. Quantum Electron.*, Vol. QE-21, 1985, p. 121.

[24] Olshansky, R., P. Hill, V. Lanzisera, and W. Powazinik, "Frequency Response of 1.3 μm InGaAsP High Speed Semiconductor Lasers," *IEEE J. Quantum Electron.*, Vol. QE-23, 1987, p. 1410.

[25] Kobayashi, S., Y. Yamamoto, M. Ito, and T. Kimura, "Direct Frequency Modulation in AlGaAs Semiconductor Lasers," *IEEE J. Quantum Electron.*, Vol. QE-18, 1982, p. 582.

[26] Yoshikuni, Y., and G. Motosugi, "Multielectrode Distributed Feed Back Laser for Pure Frequency Modulation and Chirping Suppressed Amplitude Modulation," *IEEE J. Lightwave Technol.*, Vol. LT-5, 1987, p. 516.

[27] Ishida, O., H. Toba, K. Tomori, and K. Oe, "Multielectrode DBR Laser Diode for Wide Bandwidth and Flat FM Response," *Electron. Lett.*, Vol. 25, 1989, p. 703.

[28] Schawlow, A. L., and C. H. Townes, "Infrared and Optical Masers," *Phys. Rev.*, Vol. 112, 1958, p. 1940.

[29] Henry, C. H., "Theory of the Linewidth of Semiconductor Lasers" *IEEE J. Quantum Electron.*, Vol. QE-18, 1982, p. 259.

[30] Furuya, K., Y. Suematu, and T. Hong, "Reduction of Resonance-Like Peak in Direct Modulation Due to Carrier Diffusion in Injection Laser," *Appl. Opt.*, Vol. 17, 1978, p. 1949.

[31] Nagano, K., M. Maeda, K. Sato, M. Tanaka, and R. Ito, "Sinusoidal Modulation Characteristics of Buried-Heterostructure Lasers," *Trans. IECE Japan*, Vol. E61, 1978, p. 441.

[32] Loudon, R., *The Quantum Theory of Light*, Clarendon Press, 1973.

[33] Papoulis, A., *Probability, Random Variables, and Stochastic Process*, McGraw-Hill, 1965.

[34] McIntyre, R. J., "Multiplication Noise in Uniform Avalanche Diodes," *IEEE Trans. Electron. Devices*, Vol. ED-13, 1966, p. 164.

[35] Imai, H., and T. Kaneda, "High-Speed Distributed Feedback Lasers and InGaAs Avalanche Photodiodes," *IEEE J. Lightwave Technol.*, Vol. 6, 1988, p. 1634.

[36] Taguchi, K., T. Torikai, Y. Sugimoto, K. Makita, and H. Ishihara, "Planar-Structure InP/InGaAsP/InGaAs Avalanche Photodiodes With Preferential Lateral Extended Guard Ring for 1.0–1.6 μm Wavelength Optical Communication Use," *IEEE J. Lightwave Technol.*, Vol. 6, 1988, p. 1643.

[37] Kawachi, M., K. Kobayashi, and T. Miyashita, "Hybrid Optical Integration With High-Silica Channel Waveguide on Silicon," *Proc. Top. Meet. Integrated and Guided-Wave Opt.*, 1986, p. 62.

[38] Beguin, A., T. Dumas, M. Hackert, R. Jansen, and C. Nissum, "Fabrication and Performance of Low-Loss Optical Components Made by Ion Exchange in Glass," *IEEE J. Lightwave Technol.*, Vol. 6, 1988, p. 1483.

Chapter 3
Digital and Analog Transmission

There are many digital and analog transmission technologies, for example, *intensity modulation/direct detection* (IM/DD), coherent transmission, and subcarrier multiplexing technologies. Historically speaking, IM/DD technology was the first one developed. This technology is simple and has been used for many existing optical systems. Coherent transmission technology uses the wave aspect of light positively. This technology is similar to that used for existing AM or FM radio broadcasts. Many well-known technologies, such as FSK, *phase-shift keying* (PSK), heterodyne detection, and homodyne detection, can be applied to coherent transmission technologies. By using these coherent transmission technologies, longer transmission distance without repeaters is possible. Although this feature is very important for trunk transmission, it may not matter for subscriber systems. Optical frequency selectivity, which is another important feature of coherent transmission, may be rather important for subscriber systems. Subcarrier multiplexing is the multiplexing technology that uses electrical subcarriers. Optical fiber has a very small core diameter, especially for a single-mode fiber. Both small core diameter and long fiber length result in the situation of strong light intensity and long interaction length. These situations cause several nonlinear effects in a fiber: stimulated Raman scattering, stimulated Brillouin scattering, four-wave mixing, optical Kerr effect, and so on. These interesting phenomena impose some limitations on optical transmission in addition to being used for some useful applications. In this chapter, IM/DD, coherent transmission, SCM, and fiber nonlinearity are explained.

3.1 INTENSITY MODULATION AND DIRECT DETECTION

IM/DD technology has been widely used and will be widely used for the mean time of optical subscriber systems because of its simplicity. The principle of IM/DD is shown in Figure 3.1. The intensity of laser light is either directly modulated or externally modulated by digital or analog electric signals. As a result of the modulation,

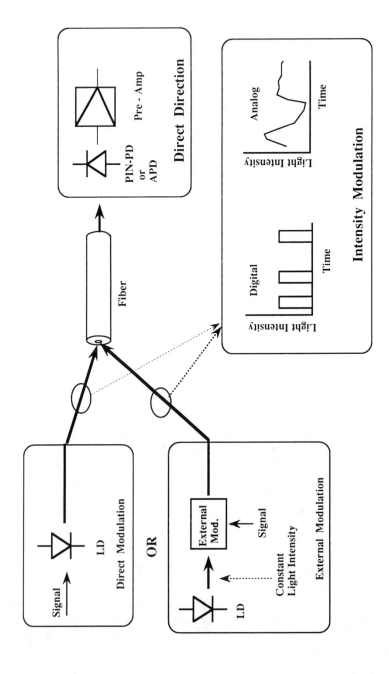

Figure 3.1 Principle of IM/DD.

light intensity varies in digital or analog format. The intensity modulated light, which is transmitted by a fiber, is directly detected by a photodiode (pin-PD or APD).

3.1.1 Intensity Modulation

Commercially available laser diodes have the threshold current I_{th}, and they emit laser light when the injected current value is above the threshold. Today, the value of the threshold current I_{th} ranges from 10 to 20 mA for commercial LDs. LDs with $I_{th} < 1$ mA were realized at the laboratory level [1], and LDs without the threshold have been investigated [2]. The following discussion also holds by setting $I_{th} = 0$ for future LDs with no threshold. The modulation methods for LDs are shown in Figures 3.2 and 3.3. Figure 3.2 corresponds to the case of the bias current $I_b < I_{th}$, and Figure 3.3 to $I_b = I_0$. I_0 is the bias current where no signal distortion is realized ($I_0 > I_{th}$).

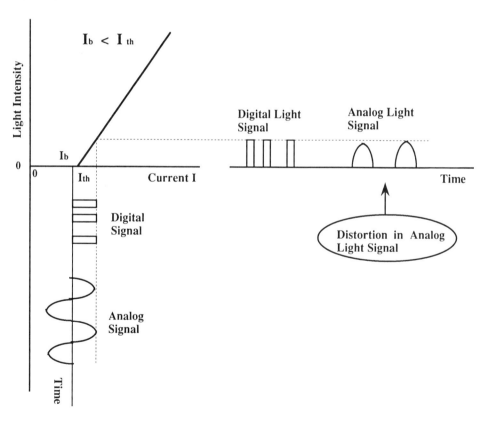

Figure 3.2 LD biasing with $I_b < I_{th}$.

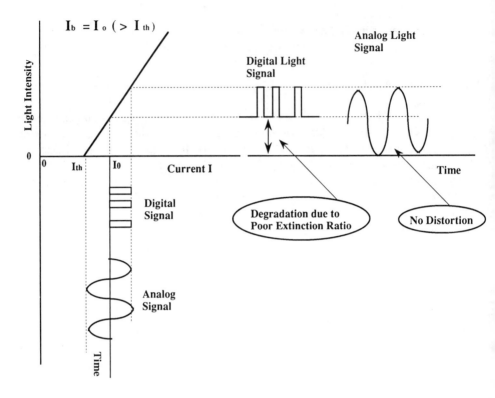

Figure 3.3 LD biasing with $I_b = I_0$.

As indicated in these figures, there exist signal distortions in light signals for $I_b < I_{th}$. Therefore, I_b must be set to I_0 for analog transmissions, where distortion is harmful to system performance. On the other hand, the condition of $I_b < I_{th}$ is recommended for digital transmissions. When I_b is set above I_{th}, light is always emitted from a laser diode. If "mark" corresponds to higher intensity level, low-level-intensity light comes into a photodiode in a receiver for a "space" period. In this case, the degradation in receiving sensitivity due to poor extinction ratio occurs, which will be discussed later in this chapter. So, the condition of $I_b < I_{th}$ is recommended for digital transmissions.

For digital transmissions, $I_b = 0$ (zero bias) is possible. Although the modulation circuit is simple in a zero bias condition, a large current amplitude change (large current swing) must be required for an electric driving circuit when compared to the nonzero bias condition. For example, the value of $I_{th} + 20$ mA current is approximately required for 1 mW (0 dBm) of output light from a single-mode fiber in a typical commercial laser diode module, in which a laser is attached to a single-mode fiber. It is known that the threshold current depends on temperature. It is expressed as

$$I_{th} = I_0 \exp(T/T_0) \tag{3.1}$$

where, I_0, T_0, and T are the constant, the specific temperature, and the ambient temperature, respectively. The value of T_0 ranges from 40° K to 50° K for ordinary InGaAsP lasers. In the above example, $I = 40$ mA for room temperature, and $I \sim 70$ mA for 60°C (= 333° K) when $I_{th} = 20$ mA. In the case of $I_b = 0$, current swing is about 70 mA at 60°C. Generally speaking, large current swing like this is difficult for high frequencies.

One example of the modulation circuit, which is commonly used, is shown in Figure 3.4. The constant current source in this figure is also realized by transistor circuits. The input signal applies the base of Tr 1 (transistor), and the base voltage V_B of Tr 2 is constant. Circuit parameters are determined that a signal circuit with a constant dc bias current I_b imposes on a laser diode. When designing a practical modulation circuit, change of ambient temperature must be taken into consideration. There are two ways to stabilize the output laser power against the ambient temperature change: (1) *Automatic power control* (APC) using a light sensor (a monitor

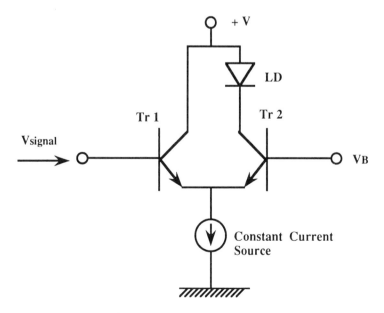

Figure 3.4 Example of LD modulation circuit.

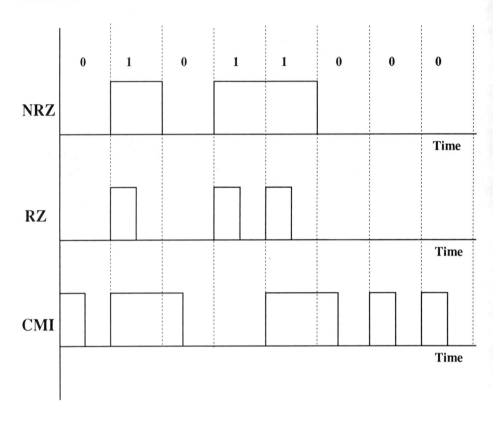

Figure 3.5 Line coding.

photodiode); and (2) *Automatic temperature control* (ATC) using a temperature sensor (thermistor).

Either of these or both can be used. In the first method, output light power is monitored by a monitor photodiode, and the detected photocurrent is used as a feedback signal to the bias current I_b. To obtain the constant output power, I_b increases as the ambient temperature increases, as indicated by (3.1). In the second method, a thermistor and a Peltier device are commonly used, the thermistor as a temperature monitor and the Peltier device as a heater or a cooler. As the temperature of a laser module is stabilized, I_b stays constant regardless of the ambient temperature change. The disadvantage in the second method, which may be disliked because of the equipment placed with the subscribers, is the relatively large current consumption. For example, about ±0.5A is required to stabilize a commercial laser module's temperature for ambient temperature changes from −20°C to +60°C. However, the second method is also useful for wavelength stabilization because the laser wavelength is changed by temperature changes. The principal origin of the wavelength change is the mode hopping for an FP-LD and the refractive index change for a DSM-LD (such as a DFB-DB or a DBR-LD).

There are many line codes, which are selected and devised for optical systems by considering the timing extraction, bandwidth, and so on. Three examples of them are shown in Figure 3.5. They are NRZ, *return to zero* (RZ), and CMI codes. These codes are frequently used in practical optical systems.

3.1.2 Direct Detection

Detection for Digital Signals

Expressions for error rate are derived here. Many authors have discussed these topics [3–6]. The derivation method of error rate is similar to the well-known method used for metallic *pulse-code modulation* (PCM) transmission. However, the intrinsic noise of lightwave systems does exist, and it is known as quantum noise (shot noise), as discussed in Chapter 2.

The noise σ^2 in an optical receiver is composed of the signal shot noise, the dark current shot noise, and the thermal (circuit) noise:

$$\sigma^2 = \langle i_s^2 \rangle + \langle i_d^2 \rangle + \langle i_t^2 \rangle \tag{3.2}$$

where $\langle i_s^2 \rangle$ is signal shot noise; $\langle i_d^2 \rangle$ is dark current shot noise; and $\langle i_t^2 \rangle$ is thermal (circuit) noise.

$\langle i_s^2 \rangle$ is (2.45) for PD and (2.49) for APD. $\langle i_d^2 \rangle$ is (2.50) for PD and (2.51) for APD. Thermal noise due to the preamplifier circuits is

$$\langle i_t^2 \rangle = \frac{4kT}{R_L} FB \tag{3.3}$$

where k, T, F, and R_L are the Boltzmann constant ($= 1.38 \times 10^{-23}$ [J/K]), kelvin temperature, the noise figure of the amplifier, and the load resistance, respectively. The noise σ^2 in an optical receiver is expressed as

$$\sigma^2 = 2e[\eta e \{(P_s/hf) + I_m\} \langle M \rangle^{2+x} + I_n] B + \frac{4kT}{R_L} FB \quad \text{(APD)} \tag{3.4}$$

$$\sigma^2 = 2e[\eta e (P_s/hf) + I_d] B + \frac{4kT}{R_L} FB \quad \text{(PD)} \tag{3.5}$$

The noises for mark (1) and space (0) are expressed by the notations σ_1 and σ_0 as

$$\sigma_1^2 = \langle i_{s1}^2 \rangle + \langle i_d^2 \rangle + \langle i_t^2 \rangle \tag{3.6}$$

$$\sigma_0^2 = \langle i_{s0}^2 \rangle + \langle i_d^2 \rangle + \langle i_t^2 \rangle \tag{3.7}$$

Here, we think of the higher level of light intensity as "mark." Signals detected by a photodiode are expressed by

$$i_s = \eta \, e \, \langle M \rangle \, (P_s/hf) \quad \text{(APD)} \tag{3.8}$$

$$i_s = \eta \, e \, (P_s/hf) \quad \text{(PD)} \tag{3.9}$$

The detected current for mark is larger than for space. Therefore, $\sigma_1 > \sigma_0$ ($\sigma_1 \neq \sigma_0$), and this means that the noise of an optical receiver depends on the input signal light level. The current level S_1 is defined as a mark level when $P_s = P_1$. The level S_0 is defined as a space level when $P_s = P_0$.

Error rate P_e depends on the probability of mark and space, and this is

$$\begin{aligned} P_e &= \text{Prob}(0) \, P_e(0) + \text{Prob}(1) \, P_e(1) \\ &= [P_e(0) + P_e(1)]/2 \end{aligned} \tag{3.10}$$

where Prob(0) is the probability of 0 and Prob(1) is that of 1. Here, these probabilities are assumed to be the same for the sake of simplicity. To calculate the error rate, the following Gaussian approximation gives a reasonably good estimate of the optical receiver sensitivity [7]:

$$\begin{aligned} P_e &= \frac{1}{2} \left[\frac{1}{\sqrt{2\pi}\,\sigma_0} \int_D^\infty e^{-(i-S_0)^2/2\sigma_0^2} \, di + \frac{1}{\sqrt{2\pi}\,\sigma_1} \int_{-\infty}^D e^{-(i-S_1)^2/2\sigma_1^2} \, di \right] \\ &= \frac{1}{2} \left[\frac{1}{\sqrt{2\pi}} \int_{Q_0}^\infty e^{-x^2/2} \, dx + \frac{1}{\sqrt{2\pi}} \int_{Q_1}^\infty e^{-x^2/2} \, dx \right] \end{aligned} \tag{3.11}$$

where

$$Q_0 = \frac{D - S_0}{\sigma_0} \tag{3.12}$$

$$Q_1 = \frac{S_1 - D}{\sigma_1} \tag{3.13}$$

D is the decision level, as shown in Figure 3.6. When we select D for $P_e(0) = P_e(1)$, then $Q_0 = Q_1 (= Q)$. In this case, the following equations hold:

$$Q = \frac{S_1 - S_0}{\sigma_1 + \sigma_0} \qquad D = \frac{\sigma_0 S_1 + \sigma_1 S_0}{\sigma_1 + \sigma_0} \tag{3.14}$$

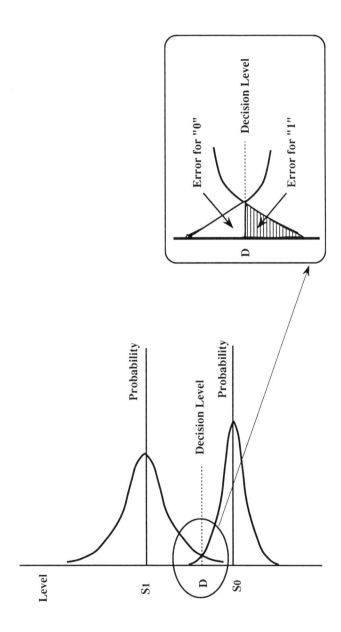

Figure 3.6 Decision level in IM/DD scheme.

$$P_e(Q) = \frac{1}{\sqrt{2\pi}} \int_Q^\infty e^{-x^2/2} \, dx = \frac{1}{2} \operatorname{erfc}\left(\frac{Q}{\sqrt{2}}\right) \tag{3.15}$$

where

$$\operatorname{erfc}(x) = \frac{2}{\sqrt{\pi}} \int_x^\infty e^{-t^2} \, dt = \frac{2}{\sqrt{2\pi}} \int_{\sqrt{2}x}^\infty e^{-y^2/2} \, dy \tag{3.16}$$

Equation (3.15) holds generally, and it can be applicable both for a receiver with a PD and for a receiver with an APD. Two approximations of (3.15) are listed below, and the calculated error rates for various Q values are listed in Table 3.1.

Table 3.1
Comparison of Strict Value and Approximation in Error Rate Calculation

Q	1	2	3	4	5	6
Strict	0.1587	0.0227	0.00135	3.16×10^{-5}	2.87×10^{-7}	9.9×10^{-10}
Approximation 1 Eq. (3.17)	0.0726	0.0223	0.00136	3.20×10^{-5}	2.89×10^{-7}	9.9×10^{-10}
Approximation 2 Eq. (3.18)	0.2420	0.0270	0.00148	3.35×10^{-5}	2.97×10^{-7}	1.0×10^{-9}

$$P_e \approx \frac{1}{Q\sqrt{2\pi}} \left(1 - \frac{0.7}{Q^2}\right) e^{-Q^2/2} \quad \text{(Approximation 1)} \tag{3.17}$$

$$P_e \approx \frac{1}{Q\sqrt{2\pi}} e^{-Q^2/2} \quad \text{(Approximation 2)} \tag{3.18}$$

The relationship between the *signal-to-noise ratio* (S/N) and the error rate can be obtained using the S/N of "mark."

$$\gamma_B = \left(\frac{S}{N}\right)_{\text{mark}} = \frac{S_1^2 R_L}{\sigma_1^2 R_L} = \frac{S_1^2}{\sigma_1^2} \tag{3.19}$$

To obtain the simple equation, the following two assumptions are used. First, no light is transmitted when the signal is space. Therefore, $S_0 = 0$. Second, although $\sigma_1 > \sigma_0$ ($\sigma_1 \neq \sigma_0$) holds, we assume $\sigma_1 = \sigma_0$.

Using these approximations, Q is related by

$$Q = \frac{S_1}{2\sigma_1} = \frac{1}{2}\sqrt{\gamma_B} \tag{3.20}$$

Therefore,

$$P_e(\gamma_B) = \frac{1}{\sqrt{2\pi}} \int_{1/2\sqrt{\gamma_B}}^{\infty} e^{-x^2/2}\, dx \tag{3.21}$$

For example, $Q = 6$ when $P_e = 10^{-9}$. Then we obtain $\gamma_B = 21.6$ dB. The calculated error rate as a function of S/N at mark is shown in Figure 3.7.

Sensitivity of receivers using PD. Sensitivity is defined as the minimum input light power of a receiver for obtaining the required error rate (usually $P_e = 10^{-9}$). Although the sensitivity is obtained by using (3.15), it can be simply expressed for a

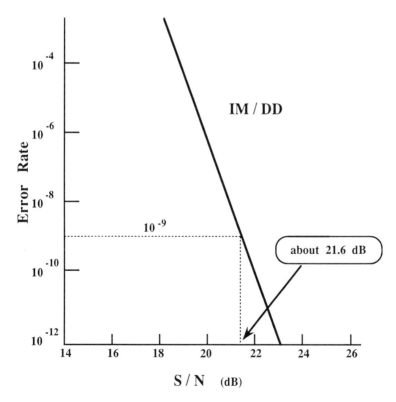

Figure 3.7 Error rate and S/N in IM/DD scheme.

receiver using a PD, which we show below. The major noise of a receiver using a PD is the thermal noise of the preamplifier in general. Therefore, $\sigma_1 = \sigma_0 = \langle i_t^2 \rangle^{1/2}$ holds in this case. The extinction ratio Ex is defined as

$$\text{Ex} \equiv \frac{P_1}{P_0} \tag{3.22}$$

Using this Ex, parameter Q is expressed as

$$Q = \frac{S_1 - S_0}{\sigma_1 + \sigma_0} = \frac{\eta e(P_1 - P_1/\text{Ex})}{2hf \langle i_t^2 \rangle^{1/2}} \tag{3.23}$$

Equation (3.23) is rewritten as

$$P_1 = \frac{2hf Q \langle i_t^2 \rangle^{0.5}}{\eta e(1 - 1/\text{Ex})} \tag{3.24}$$

The average power P is expressed using (3.24):

$$P = \frac{1}{2}(P_1 + P_0) = \frac{1}{2}\left(1 + \frac{1}{\text{Ex}}\right)P_1 = \frac{(\text{Ex} + 1)}{(\text{Ex} - 1)}\left(\frac{hf}{\eta e}\right) \langle i_t^2 \rangle^{0.5} Q \tag{3.25}$$

For Ex $\to \infty$ (extinction ratio tends to infinity), (3.25) is

$$P = \left(\frac{hf}{\eta e}\right) \langle i_t^2 \rangle^{0.5} Q \quad (\text{Ex} \to \infty) \tag{3.26}$$

When comparing (3.26) with (3.25), the degradation due to the finite Ex value is

$$\Delta P = 10 \log\left(\frac{\text{Ex} + 1}{\text{Ex} - 1}\right) \quad (\text{dB}) \tag{3.27}$$

The required input light power increases by ΔP. For example, $\Delta P = 0.87$ dB in the case of Ex $= 10$.

Receivers using APD. In the case of receivers using an APD with a large M value, the receiver sensitivity improves by decreasing the thermal circuit noise equivalently. Although the sensitivity improves, there are some gaps between this approach and the quantum limit of IM/DD, which is discussed later (see "Ideal detection for digital signals in IM/DD" below). Towards the quantum limit, coherent detection technology and the optical amplifier have been investigated and used. These are discussed later. In the case of receivers using an APD, which decreases the circuit noise equivalently in the S/N expression, the approximation of $\sigma_1^2 \approx \langle i_t^2 \rangle$ cannot be used. As discussed above, the noise of a receiver using an APD increases with $\langle M \rangle^{2+x}$ + C, where C is a constant, and it corresponds to the thermal noise. On the other hand, the signal increases with $\langle M \rangle^2$. Therefore, the S/N increases with the increase of M at the lower value of M. However, the S/N decreases with the increase of M at the higher value of M because the term $\langle M \rangle^{2+x}$ is dominant at the higher value of M. There exists the optimum M value, Mopt, for the maximum S/N.

Although the thermal noise is not the dominant noise, the following rough estimation holds using the assumption of the thermal noise being the dominant noise.

$$\eta_{APD} P_{APD} = \frac{2}{\langle M_{OPT} \rangle} \eta_{PD} P_{PD} \tag{3.28}$$

where η_{APD} and η_{pin} are quantum efficiencies for the APD and PD. Receiver sensitivities using the APD increase about 17 dB for $M_{opt} = 100$, assuming $\eta_{APD} = \eta_{pin}$. This increases about 12 dB for $M_{opt} = 30$. Muoi [8] indicates that the increase of 10 to 15 dB for a Si APD (0.8-μm wavelength range) with $M_{opt} = 30$ to 100, and 5 to 10 dB for an InGaAs APD (1.3- to 1.5-μm wavelength range) with $M_{opt} = 5$ to 35.

Precise analysis by S. D. Personick. Personick makes a precise analysis including the waveform effect of a pulse [3–5]. His results are shown below.

$$\langle i_s^2 \rangle = 2e \left(\frac{\eta e}{hf} \right) \langle M \rangle^{2+x} P_s B I_1 \tag{3.29}$$

$$\langle i_d^2 \rangle = 2e[I_m \langle M \rangle^{2+x} + I_n] B I_2 \tag{3.30}$$

For a *field-effect transistor* (FET) front-end,

$$\langle i_t^2 \rangle = \left[\frac{4kT}{R_L} \left(1 + \frac{\Gamma}{g_m R_L} \right) + 2eI_{gate} \right] BI_2 + 4kT \Gamma \frac{(2\pi C_T)^2}{g_m} B^3 I_3 \tag{3.31}$$

where

I_{gate} = gate leakage current;
g_m = transconductance;

Γ = numerical factor for FET ($\Gamma = 0.7$ for Si FET and $\Gamma = 1.1$ for GaAs FET); and

C_T = total capacitance (which is the sum of the gate-source capacitance and gate-drain capacitance, and APD (or PD) capacitance and stray capacitance).

For a bipolar transistor front-end,

$$\langle i_t^2 \rangle = 2eI_b BI_2 + \frac{2eI_c}{g_m^2}[(1/R_L + 1/r_{b'e})^2 BI_2 + (2\pi C_T)^2 B^3 I_3]$$
$$+ 4kT\, r_{b'b} \left[\frac{BI_2}{R_L^2} + 4\pi^2(C_t + C_s)^2 B^3 I_3\right] \quad (3.32)$$

where the symbols are:

I_b = base current;
I_c = collector current;
g_m = transconductance ($= eI_c/kT$);
$r_{b'b}$ = base resistance;
$r_{b'e}$ = base-emitter resistance; and
C_T = total capacitance (which is the sum of the base-emitter capacitance and base-collector capacitance, and APD (or PD) capacitance(C_t) and stray capacitance (C_s).

I_1, I_2, I_3 are constants defined by Personick, which vary according to the shape of a pulse. The typical values are:

$I_1 = 0.5$,
$I_2 = 0.4$ to 0.5, and
$I_3 = 0.03$ to 0.09.

The values for I_1 to I_3 are shown in [3,4].

Ideal detection for digital signals in IM/DD. Here, we consider the ideal case, where no circuit noise and no dark current exist, and quantum efficiency $\eta = 1$. The noise source is only the shot noise due to the signal light. The light pulse, which is composed of ΛT photons on average, is transmitted for mark, and no light is transmitted for space. Since no noise exists for space, the decision level can be set at zero. The error occurs only for mark because no noise exists for space. The error for mark is the probability of no photons for T seconds, where T is the time slot:

$$P(n = 0) = e^{-\Lambda T} \quad (3.33)$$

This is obtained by using (2.40). The error rate is the average of the probabilities for mark and space, and we assume the same probability of mark and space:

$$P_e = \frac{1}{2}(e^{-\Lambda T} + 0) = \frac{1}{2}e^{-\Lambda T} \qquad (3.34)$$

This is the limit for IM/DD and is called the *quantum limit*. For $P_e = 10^{-9}$, $\Lambda T = 20$. For the average power, $P = (20 + 0)/2 = 10$, and that is 10 photons per bit. The minimum power for R_b bit rate in IM/DD is

$$P = 10\, hfR_b \qquad (3.35)$$

For example, $P = 1.53 \times 10^{-15} \times R_b$ [mW] at the 1.3-μm wavelength. For $R_b = 500$ Mb/s, $P = -61$ dBm. The actual receiver is not ideal, so $P = -44$ dBm for the reported good IM/DD receiver at 500 Mb/s [8]. The gap from the ideal value is 17 dB. Table 3.2 shows the photon numbers and the degradation numbers for the state of the art. The sensitivity in IM/DD improves when an optical amplifier, such as an *Er-doped-fiber amplifier* (EDFA), is used. (EDFA is explained in Chapter 5.) For example, 147 photons/bit is reported at 10 Gb/s using the EDFA [9]. In this case, the gap between the *obtained value* (147 photons/bit) and the ideal value (10 photons/bit) is 11.7 dB.

Table 3.2
Receiver Sensitivity in IM/DD

	Photon/bit	Degradation (dB)
Ideal	10	0
State of the art		
InGaAs pin-PD	6,000–20,000	28–33
InGaAs APD	500–2,000	17–23
EDFA + InGaAs pin-PD	147	11.7

Detection for Analog Baseband Signals

Detection for optical analog signals is also similar to that for ordinary electric signals, apart from the quantum noise.

General case for analog signals in IM/DD. The modulated optical signals are expressed as

$$P(t) = P_0[1 + m(t)] \qquad (3.36)$$

where $m(t)$ is the modulated analog signal (AM, FM, etc.) with bandwidth B and $|m(t)| \leq 1$ and $\langle m(t) \rangle = 0$. For example, the amplitude of $m(t)$ is modulated for AM and the frequency of $m(t)$ is modulated for FM. P_0 is the average incident optical power. The received carrier is

$$i = \eta e(P_0/hf) \tag{3.37}$$

Then the *carrier-to-noise ratio* (C/N) is

$$\frac{C}{N} = \frac{i^2}{\sigma^2} \tag{3.38}$$

The relation between C/N and S/N depends on the modulation scheme, which is discussed in many books, such as [10].

Ideal detection for analog signals in IM/DD. Here, we consider the ideal case, where no circuit noise and no dark current noise exist. In this ideal case, the only noise source is shot noise due to the signal light. The received carrier and noise are

$$i = \eta e(P_0/hf) \tag{3.39}$$

$$\sigma^2 = \langle i_s^2 \rangle = 2e[\eta e(P_0/hf)]B \tag{3.40}$$

Then the C/N is

$$\frac{C}{N} = \frac{i^2 R_L}{\sigma^2 R_L} = \frac{\eta P_0}{2hfB} \tag{3.41}$$

When the required C/N is 50 dB for AM with $B = 6$ MHz, $P_0 = -37.4$ dBm for the 1.3-μm wavelength and $\eta = 1$.

3.2 COHERENT TRANSMISSION

Coherent transmission technology [11–13] uses the wave aspect of light, while IM/DD uses the intensity of light. There are three major advantages with using optical coherent technology, such as the receiver sensitivity improvement, the good optical frequency selectivity, and the possibility of equalization at the *intermediate frequency* (IF) band, when compared to IM/DD. Using the superior optical frequency selectivity, densely spaced WDM is realized with the optical coherent technology, which is known as *optical frequency-division multiplexing* (OFDM). The wavelength spacing of ordinary WDM is about 100 to 300 nm, and that of OFDM is about 0.08

nm when $\Delta f = 10$ GHz. In this book, OFDM using IM/DD is classified as coherent technology because it uses the wave aspect of light [14]. The multiplexing or selection of an optical frequency in OFDM is usually done by using interferometers, such as the *Mach-Zehnder* (MZ) interferometer and the *Fabry-Perot* (FP) interferometer. These are explained in Chapter 8. The interferometer is based on the wave aspect of light. An example of equalization at the IF band is the compensation of fiber chromatic dispersion by an electric equalizer [15].

3.2.1 Principle

The optical coherent transmission scheme is similar to that used in existing radio wave transmissions, shown in Figure 3.8. In both cases, the received signals are mixed with the waves from the local oscillator in a receiver. Two waves, signal and

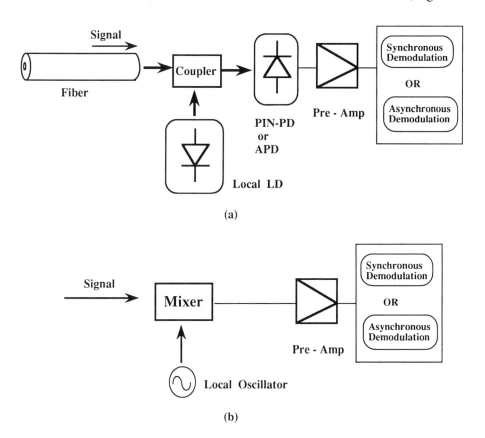

Figure 3.8 Coherent transmission: (a) optical coherent transmission; (b) radio wave transmission.

local oscillator waves, are mixed in coherently. Signals are detected synchronously or asynchronously from the mixed waves in electrical processing. Asynchronous detection is envelope detection, and synchronous detection uses the reference electrical wave, which is synchronous with the signal electric wave. In optical coherent transmission, a PD (or an APD) is used both as an optical detection device and as a mixing device. In microwave transmission, several modulation schemes have been developed and used. For digital signals, *amplitude shift keying* (ASK), FSK, PSK, and *amplitude phase shift keying* (APSK) are typical schemes. The amplitude, frequency, and phase parameters are modulated in ASK, FSK, and PSK, respectively. Both amplitude and phase are modulated in APSK. These are shown in Figure 3.9. *Continuous phase FSK* (CPFSK) has the characteristic of continuous phase of the carrier, and it results in good spectrum concentration. *Minimum shift keying* (MSK) and *Gaussian-filtered MSK* (GMSK) are two examples of CPFSK. One example of APSK is *quadrature amplitude modulation* (QAM), and examples of this are 16 QAM, 64 QAM, and 256 QAM. In *differential phase-shift keying* (DPSK), which belongs to PSK, differential coding is accomplished by the phase change of sequential pulses. Some of these schemes have been investigated and used for optical coherent transmission.

Homodyne detection and heterodyne detection have been investigated and are shown in Figure 3.10. Both the optical frequency and the phase of the local oscillator wave are identical and locked to those of the input signal wave in homodyne detection. In heterodyne detection, the frequency of the local oscillator wave is different from that of the input signal wave. The signal wave is expressed as

$$S(t) = A \cos(\omega_s t + \varphi_s) \tag{3.42}$$

and the local oscillation wave is

$$L(t) = B \cos(\omega_L t + \varphi_L) \tag{3.43}$$

where A and B are the amplitudes and ω_s and ω_L are the angular frequencies, and φ_s and φ_L are the phases. When the balanced mixer detection is used (discussed later), the instantaneous input power of an APD (or a PD) is

$$\begin{aligned}P_{in} &= [A \cos(\omega_s t + \varphi_s) + B \cos(\omega_L t + \varphi_L)]^2 \\ &= \frac{1}{2}(A^2 + B^2) + AB \cos[(\omega_s - \omega_L)t + (\varphi_s - \varphi_L)] + AB \cos[(\omega_s + \omega_L)t + (\varphi_s + \varphi_L)] \\ &\quad + \frac{1}{2}A^2 \cos(2\omega_s t + 2\varphi_s) + \frac{1}{2}B^2 \cos(2\omega_L t + 2\varphi_L)\end{aligned} \tag{3.44}$$

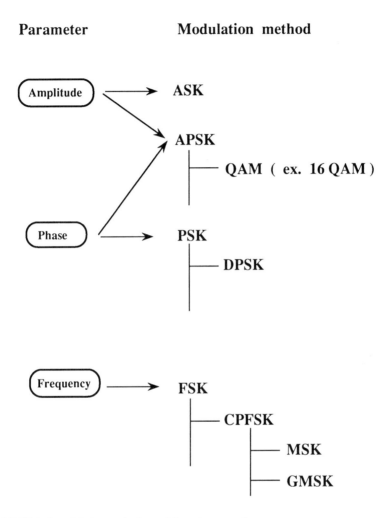

Figure 3.9 Digital modulation methods used for microwave frequency.

An APD or a PD cannot detect the fast change of signals such as a light frequency. Therefore, the output current from an APD (or a PD) is

$$i = RM \left\{ \frac{1}{2}(A^2 + B^2) + AB \cos[(\omega_s - \omega_L)t + (\varphi_s - \varphi_L)] \right\} \quad (3.45)$$

where R and M are the responsivity and the avalanche gain ($M = 1$ for a PD), respectively.

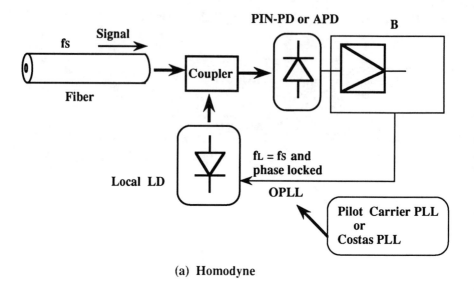

(a) Homodyne

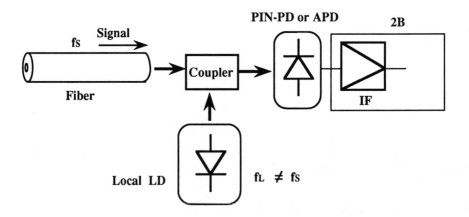

(b) Heterodyne

Figure 3.10 Homodyne and heterodyne detection.

Here, we are using M instead of $\langle M \rangle$ for the sake of simplicity (i.e., M means $\langle M \rangle$). We define the average powers as P_s and P_L for $S(t)$ and $L(t)$, respectively. Then $P_s = A^2/2$, $P_L = B^2/2$, and $P_L \gg P_s$ hold in general. Using the average powers, (3.45) is rewritten as

$$i = RM\{P_s + P_L + 2\sqrt{P_s P_L} \cos[(\omega_s - \omega_L)t + (\varphi_s - \varphi_L)]\} \tag{3.46}$$

Since $\omega_s = \omega_L$ in homodyne detection, (3.46) is

$$i = RM\{P_s + P_L + 2\sqrt{P_s P_L} \cos(\varphi_s - \varphi_L)\} \tag{3.47}$$

For homodyne detection of ASK signal with $\varphi_s = \varphi_L$, the received signal is

$$i_s = 2RM\sqrt{P_s P_L} \tag{3.48}$$

For homodyne detection of PSK signal, the received signal is

$$i_s = \pm 2RM\sqrt{P_s P_L} \tag{3.49}$$

for $\varphi_s - \varphi_L = 0$ and π.

From the derivations and results of equations (3.48) and (3.49), it is clear that the phase as well as the frequency of local oscillator wave must be locked to the signal optical wave, and the bandwidth required for the circuit is B, when the input signal bandwidth equals B. The locking is done by an *optical phase-locked loop* (OPLL) technique, and pilot carrier *phase-locked loop* (PLL) and Costas PLL are examples [16]. As for heterodyne detection, $\omega_s \neq \omega_L$ holds. The intermediate frequency (IF) is defined as

$$\omega_{IF} = \omega_s - \omega_L \tag{3.50}$$

Equation (3.46) is rewritten with ω_{IF} and $\theta = \varphi_s - \varphi_L$:

$$i = RM\{P_s + P_L + 2\sqrt{P_s P_L} \cos(\omega_{IF}t + \theta)\} \tag{3.51}$$

The received signal is

$$i_s = 2RM\sqrt{P_s P_L} \cos(\omega_{IF}t + \theta) \tag{3.52}$$

The P_s, ω_s, and φ_s parameters are changed in ASK, FSK, and PSK, respectively. The bandwidth required for circuits in heterodyne detection is $2B$, while that in homodyne detection is B.

Although single-photodiode detection is shown in Figure 3.10, the dual-photodiode approach in a balanced mixer configuration is often taken, shown in Figure 3.11. In a balanced mixer type, two beat signals from the outputs of two coupler arms are added constructively, and the local oscillator intensity noises detected at two photodiodes are added destructively (cancel) [17]. The local oscillator intensity noise originates the intensity fluctuation of local laser output. The balanced mixer configuration has the advantages both of local laser intensity noise suppression and of efficient light power usage. The calculations in (3.44) through (3.52) are assumed to be using balanced mixer detection. In the case of single-photodiode detection, P_s and P_L in these equations are replaced by xP_s and $(1 - x)P_L$, respectively. The symbol x is the coupling ratio in a coupler and $x = 0.5$ for a 3-dB coupler. In the case of a 3-dB coupler ($x = 0.5$), the received signals i_s in (3.48) through (3.52) are half (3 dB of degradation) when compared to the signals detected by a balanced mixer configuration.

The stability of the polarization state of the signal wave is required at the receiving end of the fiber. It is required for both homodyne and heterodyne detection for the effective mixing of signal light with the local laser. Four countermeasures have been considered for acquiring stability [18,19,16]. They are:

1. Use of a polarization-maintaining fiber;
2. Use of polarization state control device at receivers;
3. Use of polarization diversity receivers;
4. Use of polarization scrambling technique.

The second and third approaches are the techniques for the receiver. One example of the polarization diversity receiver is shown in Figure 3.12. Two polarization components of incident signal waves are separated by the polarization beam splitter and are individually detected by the balanced mixers. After the detection, two detected signals are summed. The principle of this approach is similar to that of the diversity for horizontal and vertical polarization waves in microwave. Although two detected signals are summed in Figure 3.12, several modifications are possible, as in the case of microwave systems. For example, only one signal that is stronger in amplitude is selectively used instead of summing. The use of the local laser and this polarization problem make the coherent transmission technology using a local oscillation laser less attractive in subscriber loop applications.

3.2.2 Receiver Configuration and Error Rate

The receiver configuration for heterodyne detection is shown in Figure 3.13 [20,21,10]. Optical signals are detected either by balanced mixer detection or single-photodiode detection. The converted IF signals are amplified and equalized if necessary. They are demodulated synchronously or asynchronously. The demodulated process is the same process as the one used in ordinary electric systems [10]. The demodulation

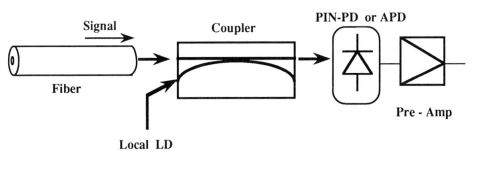

(a)

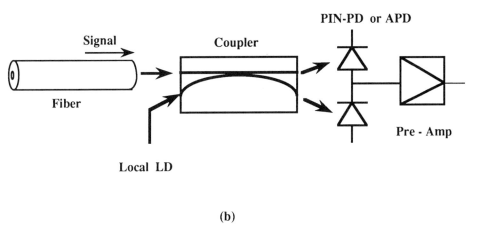

(b)

Figure 3.11 Two types of detection: (a) single-photodiode detection; (b) balanced-mixer detection.

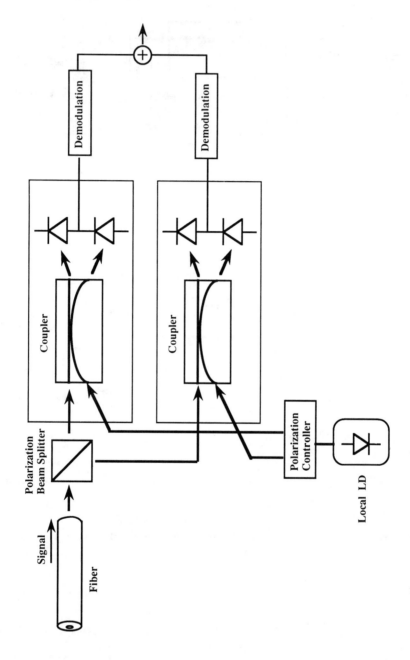

Figure 3.12 Configuration of polarization diversity receiver.

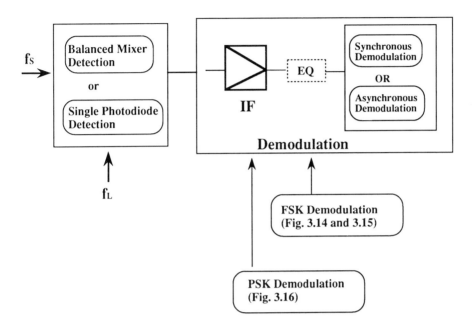

Figure 3.13 Receiver configuration for heterodyne detection.

for FSK and PSK are shown in Figures 3.14 to 3.16. In synchronous demodulation, the local electric wave is mixed with the signal electric wave. FSK synchronous demodulation uses band-pass filtering and mixing with the local electric wave, as shown in Figure 3.14. When signals with f_1 frequency transmit, the signals pass through the BPF with an f_1 center frequency and a mixing device and a *low-pass filter* (LPF), while only noise passes through the BPF with f_2. Signals and noise are compared with a *comparator* (COMP), and signals are selected. For FSK asynchronous demodulation, two possible configurations are shown in Figure 3.15: (1) envelope detection with single or dual filters and (2) detection using a discriminator. The PSK synchronous and asynchronous demodulations are shown in Figure 3.16. In synchronous demodulations, carriers for mixing are recovered from the signals by the carrier recovery circuit, such as an electric PLL circuit. In DPSK asynchronous demodulation, the delay circuit is used and the delayed signal and the original signal are mixed.

The basic receiver configuration for homodyne detection is shown in Figure 3.10(a), which uses OPLL. Other possible configurations are the injection locking and phase diversity technologies [16]. The former is the injection locking of a local laser to the signal lightwave. In the latter, the optical frequency of a local laser

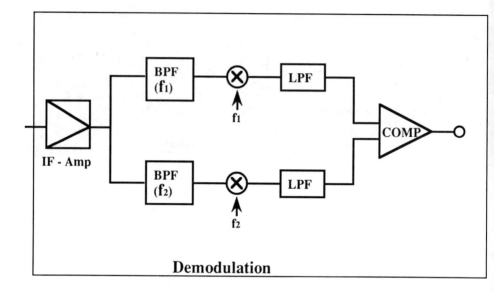

Figure 3.14 FSK synchronous demodulation.

coincides with that of a signal lightwave; however, a local laser is not phase-locked to a signal lightwave.

The error rate depends on the detection and demodulation schemes. Calculations for typical schemes are shown below. First, the signal and noise are obtained. As already discussed, the signals are

$$i_s = 2RM\sqrt{P_s P_L} \cos(\omega_{IF} t + \theta) = K \cos(\omega_{IF} t + \theta) \quad \text{(Heterodyne)} \quad (3.53)$$

$$(K \equiv 2RM\sqrt{P_s P_L})$$

$$i_s = 2RM\sqrt{P_s P_L} = K \quad \text{(ASK Homodyne)} \quad (3.54)$$

$$i_s = \pm 2RM\sqrt{P_s P_L} = \pm K \quad \text{(PSK Homodyne)} \quad (3.55)$$

$$\left(R = \frac{\eta e}{hf}\right)$$

Noise is obtained by the summation of the noises in (3.2) and the local laser shot noise.

$$\sigma^2 = \langle i_s^2 \rangle + \langle i_d^2 \rangle + \langle i_L^2 \rangle + \langle i_t^2 \rangle \quad (3.56)$$

where i_L corresponds to the local laser.

In general, $P_L \gg P_s$ holds and then

$$\sigma^2 \approx \langle i_L^2 \rangle + \langle i_t^2 \rangle + \langle i_d^2 \rangle \quad (3.57)$$

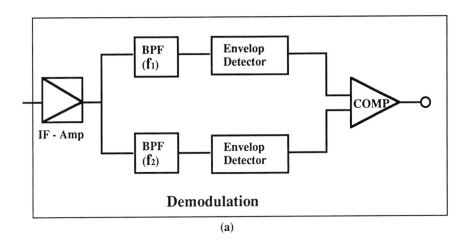

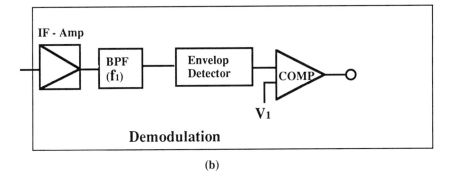

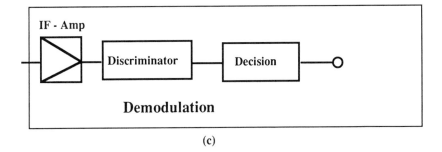

Figure 3.15 FSK asynchronous demodulation: (a) envelope detection with dual filters; (b) envelope detection with single filters; (c) detection using discriminator.

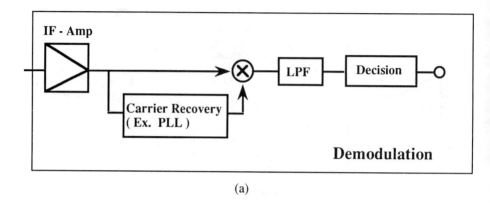

(a)

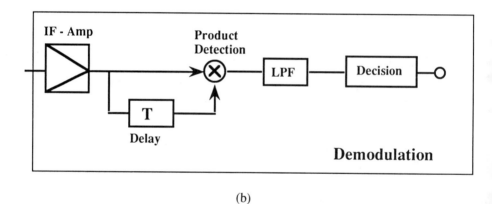

(b)

Figure 3.16 PSK synchronous and asynchronous demodulation: (a) PSK synchronous demodulation; (b) DPSK asynchronous demodulation.

In this case, $\sigma_1 = \sigma_0$ holds. The noise components except for the local laser light are shown in (3.4) and (3.5). The component for the local laser light is

$$\langle i_L^2 \rangle = 2eR\, P_L\, (2B) \quad \text{(PD, Heterodyne)} \tag{3.58}$$

$$\langle i_L^2 \rangle = 2eR\, P_L\, (B) \quad \text{(PD, Homodyne)} \tag{3.59}$$

$$\langle i_L^2 \rangle = 2eR\, P_L\, \langle M \rangle^{2+x}(2B) = 2eR\, P_L\, M^{2+x}\, (2B) \quad \text{(APD, Heterodyne)} \tag{3.60}$$

$$\langle i_L^2 \rangle = 2eR\, P_L\, \langle M \rangle^{2+x}(B) = 2eR\, P_L\, M^{2+x}(B) \quad \text{(APD, Homodyne)} \tag{3.61}$$

Here, we use M instead of $\langle M \rangle$ for the sake of simplicity (i.e., M means $\langle M \rangle$).

ASK Heterodyne Synchronous

In the case of the heterodyne detection and synchronous demodulation of optical ASK signals, the calculations are similar to those in the IM/DD.
The decision level and Q value are

$$D = \frac{S_0 + S_1}{2} \qquad Q_0 = Q_1 = Q = \frac{S_1 - D}{\sigma} = \frac{S_1}{2\sigma} \tag{3.62}$$

where $S_0 = 0$ and $S_1 = 2RM\sqrt{P_s P_L} = K$.
The error rate is

$$\begin{aligned}
P_e &= \frac{1}{2}\left[\frac{1}{\sqrt{2\pi}\,\sigma}\int_D^\infty e^{-(i-S_0)^2/2\sigma^2}\,di + \frac{1}{\sqrt{2\pi}\,\sigma}\int_{-\infty}^D e^{-(i-S_1)^2/2\sigma^2}\,di\right] \\
&= \frac{1}{2}\mathrm{erfc}\!\left(\frac{Q}{\sqrt{2}}\right) = \frac{1}{2}\mathrm{erfc}\!\left(\frac{S_1}{2\sqrt{2}\sigma}\right) = \frac{1}{2}\mathrm{erfc}\!\left(\frac{1}{2}\sqrt{\gamma}\right)
\end{aligned} \tag{3.63}$$

where

$$\gamma \equiv \frac{S_1^2}{2\sigma^2} = \frac{K^2}{2\sigma^2} \tag{3.64}$$

ASK Heterodyne Asynchronous

It is known that the output from the envelope demodulation follows the Rice distribution [10]. Equation (3.62) holds in this case.

$$\begin{aligned}
P_e &= \frac{1}{2}\left[1 - \int_D^\infty \frac{x}{\sigma^2} I_0\!\left(\frac{S_1 x}{\sigma^2}\right)\exp\!\left(-\frac{x^2 + S_1^2}{2\sigma^2}\right)dx \right. \\
&\quad \left. + \frac{1}{2}\int_D^\infty \frac{y}{\sigma^2}\exp\!\left(-\frac{y^2}{2\sigma^2}\right)dy\right] \\
&\approx \frac{1}{2}\exp\!\left(-\frac{K^2}{8\sigma^2}\right) = \frac{1}{2}\exp\!\left(-\frac{1}{4}\gamma\right)
\end{aligned} \tag{3.65}$$

The approximation in (3.65) holds when $\gamma \gg 1$.

FSK Heterodyne Synchronous

An error occurs when the noise level is higher than the signal level. In this case, the following equations hold:

$$S_0 = 0 \quad \text{and} \quad S_1 = 2RM\sqrt{P_s P_L} = K \qquad (3.66)$$

$$\begin{aligned}
P_e &= \text{Prob}(y > x) \\
&= \int_0^\infty \frac{1}{\sqrt{2\pi}\,\sigma} \exp\left(-\frac{(S_1 - x)^2}{2\sigma^2}\right) \left[\int_{y=x}^\infty \frac{1}{\sqrt{2\pi}\,\sigma} \exp\left(-\frac{y^2}{2\sigma^2}\right) dy\right] dx \qquad (3.67) \\
&= \frac{1}{2}\,\text{erfc}\left(\frac{S_1}{2\sigma}\right) = \frac{1}{2}\,\text{erfc}\left(\frac{K}{2\sigma}\right) = \frac{1}{2}\,\text{erfc}\left(\sqrt{\frac{\gamma}{2}}\right)
\end{aligned}$$

FSK Heterodyne Asynchronous

In this case, the following equations hold.

$$S_0 = 0 \qquad S_1 = 2RM\sqrt{P_s P_L} = K \qquad (3.68)$$

$$\begin{aligned}
P_e &= \text{Prob}(y > x) \\
&= \int_0^\infty \frac{x}{\sigma^2} I_0\left(\frac{S_1 x}{\sigma^2}\right) \exp\left(-\frac{x^2 + S_1^2}{2\sigma^2}\right) \left[\int_{y=x}^\infty \frac{y}{\sigma^2} \exp\left(-\frac{y^2}{2\sigma^2}\right) dy\right] dx \qquad (3.69) \\
&= \frac{1}{2}\exp\left(-\frac{K^2}{4\sigma^2}\right) = \frac{1}{2}\exp\left(-\frac{1}{2}\gamma\right)
\end{aligned}$$

PSK Heterodyne Synchronous

The signals in this case are

$$S_0 = -K \quad \text{and} \quad S_1 = K \qquad (3.70)$$

Then,

$$D = 0 \quad \text{and} \quad Q = \frac{K}{\sigma} \qquad (3.71)$$

The error rate is obtained with calculations similar to those of (3.63):

$$P_e = \frac{1}{2}\mathrm{erfc}\left(\frac{Q}{\sqrt{2}}\right) = \frac{1}{2}\mathrm{erfc}\left(\frac{K}{\sqrt{2\sigma}}\right) = \frac{1}{2}\mathrm{erfc}(\sqrt{\gamma}) \qquad (3.72)$$

ASK Homodyne

In this case,

$$S_0 = 0 \qquad S_1 = 2RM\sqrt{P_s P_L} = K \qquad (3.73)$$

$$D = \frac{K}{2} \qquad (3.74)$$

Since the bandwidth is half what it is in heterodyne detection (B instead of $2B$), noise is half. Therefore, γ in (3.63) is replaced by 2γ, and the result is

$$P_e = \frac{1}{2}\mathrm{erfc}\left(\sqrt{\frac{\gamma}{2}}\right) \qquad (3.75)$$

PSK Homodyne

In this case,

$$S_0 = -K \qquad S_1 = K \qquad D = 0 \qquad (3.76)$$

Since noise is half that which is the case in heterodyne detection, γ in (3.72) is replaced by 2γ and the result is

$$P_e = \frac{1}{2}\mathrm{erfc}(\sqrt{2\gamma}) \qquad (3.77)$$

The derived equations are expressed in γ, which corresponds to mark. In the schemes, except for ASK, the powers for mark and space are the same. In ASK, the power for space is zero. Therefore, these equations are compared in terms of the mark power. When we compare them in terms of the average power, the required average powers for signals with equal probability of mark and space in ASK is 3 dB lower than the power predicted by (3.63), (3.65), and (3.75).

Using the derived equations for error rate, the minimum required power for the 10^{-9} error rate in the ideal condition can be derived. The ideal case is known as the *quantum limit* (shot noise limit). Here, we only derived for the case of FSK heterodyne asynchronous detection. Calculations in other cases are similar. In the ideal case, the noise is the shot noise from the local laser light and the signal light ($P_L \gg P_s$ holds, then $\langle i_L^2 \rangle \gg \langle i_S^2 \rangle$). $M = 1$ (PD) and $\eta = 1$ are also assumed in this case. Using the assumptions, the following equations hold:

$$\sigma^2 \approx \langle i_L^2 \rangle = 4eRP_L B \tag{3.78}$$

Since $K^2 \equiv 4R^2 P_s P_L$ is defined, then

$$\gamma = \frac{K^2}{2\sigma^2} = \frac{RP_s}{2eB} = \frac{P_s}{2hfB} \tag{3.79}$$

The error rate in this ideal case is

$$P_e = \frac{1}{2} \exp\left(-\frac{1}{2}\gamma\right) \quad \text{(quantum limit)} \tag{3.80}$$

The minimum required photons per bit is obtained as follows. $P_e = 10^{-9}$, and then $\gamma = 40$ is obtained with (3.80). From (3.79),

$$P_s = 80hfB \quad \text{then} \quad P_s/(2B) = 40hf \tag{3.81}$$

When we assume the $2B$ bit/s for the transmission line with bandwidth B (Nyquist limit), then 40 photons/bit is required. In the ideal case, the minimum required photons per bit for several coherent transmission schemes is listed in Table 3.3 and the receiver sensitivity is shown in Figure 3.17. When we compare the sensitivity in terms of the average power, the sensitivities in ASK differ by 3 dB. They are also shown in Table 3.3 and Figure 3.17. Among coherent transmission schemes, PSK homodyne detection is the best method for the ideal case.

In the above discussion, the zero laser linewidth is assumed. However, the laser linewidth does exist in an actual laser, as discussed in the previous chapter. The required laser linewidth $\Delta\nu$ for a 1-dB penalty in receiver sensitivity is summarized in [21]. They are approximately expressed as

$$\Delta\nu < 5 \times 10^{-4} R_b \quad \text{(Homodyne)} \tag{3.82}$$

$$\Delta\nu < 3 \times 10^{-3} R_b \quad \text{(Heterodyne Synchronous PSK)} \tag{3.83}$$

Table 3.3
Ideal Receiver Sensitivity (at 10^{-9} error rate)

Modulation	Detection	
	Homodyne (photon/bit)	Heterodyne (photon/bit) *
ASK	40 (20) †	~80 (~40) †
FSK	—	~40
PSK	9	≈20

* Synchronous and asynchronous demodulation is nearly equal. (Strictly speaking, a slightly small photon number (10% less) is required in synchronous demodulation.)
† The average photon number per bit (comparison in terms of the same average power).

$$\Delta \nu < 5 \times 10^{-3} R_b \quad \text{(Heterodyne Asynchronous PSK)} \quad (3.84)$$

$$\Delta \nu < 0.1 R_b \quad \text{(Heterodyne Asynchronous FSK and ASK)} \quad (3.85)$$

where R_b is the bit rate. As indicated by the above requirement, any sort of practical homodyne detection is still a very difficult proposition, given the current stage of the technology.

The results of 78 photons/bit and 132 photons/bit were obtained in a PSK heterodyne transmission experiment for 560 Mb/s and 1.2 Gb/s using balanced mixer detection, respectively [22]. Degradation from the shot noise limit (quantum limit) is only about 6 dB in the case of 78 photons/bit. It was estimated that the degradation stemmed from the quantum efficiency of a PD (1.6 dB), thermal noise (0.7 dB), laser phase noise (0.4 dB), and electrical imperfections in the amplifiers and filters (3 dB) [22].

3.2.3 OFDM

Using superior optical frequency selectivity, the densely spaced WDM known as OFDM is possible [23,18,24,25,14]. Possible configurations for OFDM are shown in Figure 3.18.

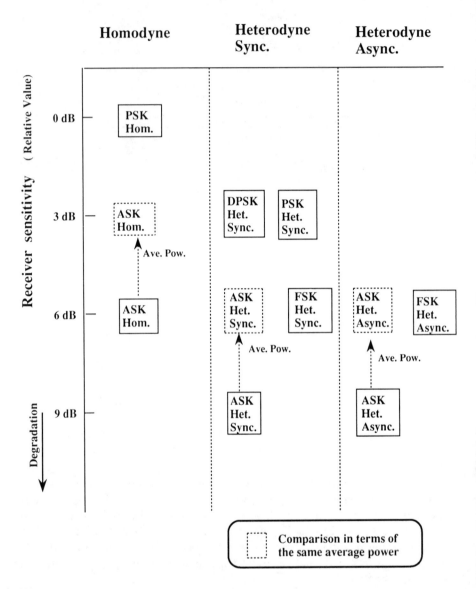

Figure 3.17 Comparison of ideal detection schemes in coherent transmission.

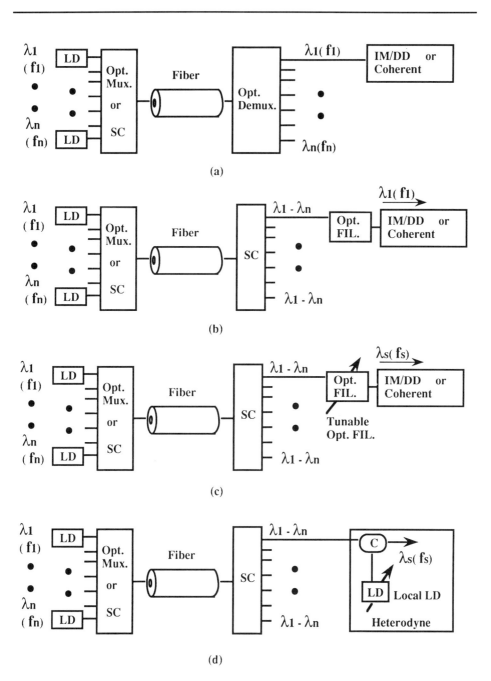

Figure 3.18 OFDM configuration: (a) method 1; (b) method 2; (c) method 3; (d) method 4.

An optical multiplexer (Opt. Mux.) or star coupler (SC) is used for combining many lights, and the combined lights transmit the same fiber. For branching the combined lights, an optical demultiplexer (Opt. Demux.) or SC is used. Here, we use the term "optical multiplexer (demultiplexer)" as the wavelength-selective device, while the "star coupler" is the wavelength-independent device (power divider or combiner). Examples of these devices are discussed in Chapter 2. The configurations in methods 1 and 2 use the dedicated wavelength (or frequency) for the receiver, and the individual receiver receives the unique wavelength, which is fixed. It is possible to receive all the wavelengths (or frequencies) in the configurations in methods 3 and 4, and the desired wavelength (or frequency) is selectively received by a tunable optical filter or a heterodyne technique. A fixed optical filter is used in method 2, while a tunable optical filter is used in method 3. For a fixed or tunable optical filter, the MZ-interferometer-type filter, the FP-interferometer-type filter, and the LD filter are used. The LD filter uses the LD amplifier characteristic, which has the narrow pass wavelength range. In the configurations of methods 1 to 3, the IM/DD transmission, as well as the coherent transmission, is possible. The frequency stability of LDs is required for all configurations. In method 4, all the waves are mixed with the local laser (LD) using a coupler, and the desired wave is selected through heterodyne detection.

The permissible channel spacing D in the case of balanced mixer detection was obtained by considering the channel interference in [26]. In [26], the following results for SIR = 30 dB (SIR is *signal-to-interference ratio*) are obtained:

$D = 3.8R_b$ (Heterodyne, FSK)
$D = 9.5R_b$ (Heterodyne, ASK)
$D = 12.4R_b$ (Heterodyne, PSK)
$D = 7.5R_b$ (Homodyne, ASK)
$D = 10.5R_b$ (Homodyne, PSK)

where R_b is bit rate. Roughly speaking, D is about $10R_b$ and very small when compared to ordinary WDM. It was pointed out that the interferences originated from the following factors [26]:

1. Intermodulation;
2. Excess shot noise by other channel;
3. Crosstalk generated by nonlinear effect in a single-mode fiber.

The nonlinear effect in a fiber is explained in Section 3.4.

3.2.4 Application in Subscriber Loops

Coherent transmission has three advantages, listed in Table 3.4 with the typical applications for trunk and subscriber systems. The major applications for subscriber loops are increasing the branching number in PON and increasing the channel number in video distribution.

Table 3.4
Advantage of Coherent Scheme and its Application

Advantage	Results	Examples of Application	
		Subscriber	*Trunk*
Receiver sensitivity	Increase of loss budget	Increase of branching number in PON	Long repeaterless transmission
Optical frequency selectivity (wavelength selectivity)	Increase of multiplexing number in WDM	Multi-channel video distribution	Increase of channel capacity Optical multiplexing
Equalization at IF	Compensation of pulse distortion	—	Long-distance transmission (compensation of fiber dispersion)

3.3 SUBCARRIER MULTIPLEXING

Transmission technologies using IM/DD and optical coherent schemes are explained in the previous sections. When we want to transmit multichannel signals efficiently, SCM is very attractive in subscriber loop applications. Digital or analog signals are multiplexed by a subcarrier and up-converted. These up-converted signals are combined and aligned in a frequency domain (subcarrier multiplexing). These subcarrier multiplexing signals (multichannel signals) are transmitted in either an IM/DD scheme or a coherent scheme. The configurations of SCM are shown in Figures 3.19 and 3.20 [27–30]. In Figure 3.19, the intensity-modulated light from SCM signals is detected by a direct-detection scheme. The detected SCM signals (n-channel signals) are multiplied by the tunable electric local oscillation, and the desired channel is picked up and demodulated. Optical coherent technology can be applicable to SCM systems, and its configurations are shown in Figure 3.20. The coherently modulated light of a laser diode is transmitted through a fiber and is detected with a local LD. The desired signal among the detected SCM signals is picked up and demodulated in the same process shown in Figure 3.19.

Systems using SCM have several advantages for multichannel transmission. It is very efficient because only one laser for multichannel transmission is required. In OFDM, a transmitting laser is not shared and each channel requires each laser. Moreover, the frequency of these lasers must be stabilized. Commercially available electronics can be used in SCM systems, and this results in a cost-effective system. Both digital (ASK, FSK, PSK) and analog (AM, FM, *phase modulation* (PM)) electric signals can be transmitted in one system (e.g., digital signals for channels 1 to 8 and analog for channels 9 to 10). The bandwidth of these signals is flexible. The

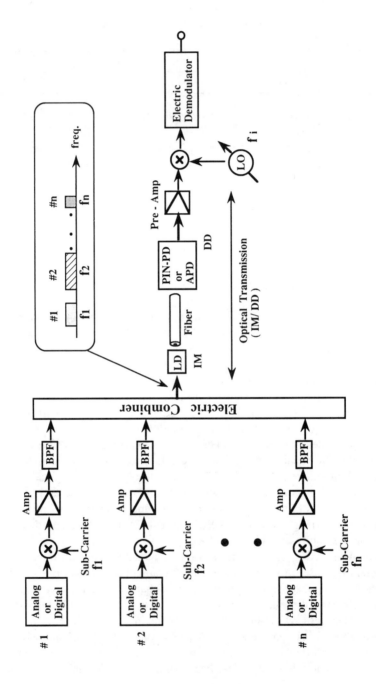

Figure 3.19 SCM system using IM/DD.

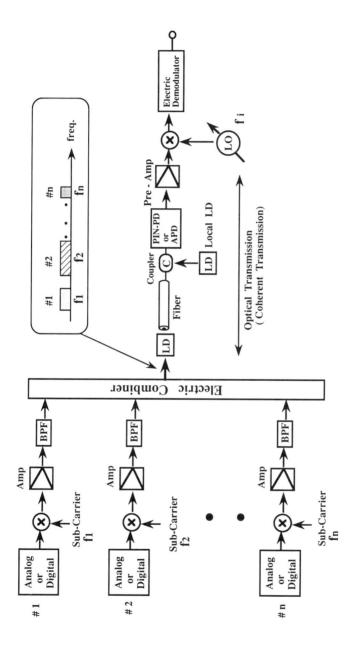

Figure 3.20 SCM system using optical coherent technology.

drawback of SCM systems is that the allowable transmission loss is small because one laser light power is shared among n channels, and some portion of the laser light (not the whole laser light) carries one channel signal.

3.3.1 SCM Systems Using IM/DD

We briefly studied the detection of analog baseband signals in Section 3.1, where the C/N is given by (3.38). In SCM, modulated optical signals are expressed as follows:

$$P(t) = P_0 \left[1 + \sum_{i=1}^{n} m_i \cos(\omega_i t + \varphi_i) \right] \tag{3.86}$$

where P_0 and n are the average received power and the number of channels, respectively. Symbols m_i, ω_i, and φ_i are the *optical modulation index* (OMI) per channel, the angular frequency, and the phase of the ith channel, respectively. Equation (3.86) in SCM systems corresponds to (3.36) in baseband systems. The output currents from an APD (or pin-PD for $M = 1$) are

$$I_{dc} = \eta e M (P_0/hf) = RMP_0 = MI_p \tag{3.87}$$

$$i_{ac} = RMP_0 \, m_i \cos(\omega_i t + \varphi_i) = m_i I_{dc} \cos(\omega_i t + \varphi_i) \tag{3.88}$$

where the notations in (3.87) are the same as those in (2.37). For the sake of simplicity, we use here M instead of $\langle M \rangle$ (M means $\langle M \rangle$). I_{dc} is the mean current (dc component) and i_{ac} is the signal ac current of channel i. The mean square current of channel i is

$$i_{ac}^2 = \frac{1}{2} m_i^2 I_{dc}^2 = \frac{1}{2} m^2 I_{dc}^2 = \frac{1}{2} m^2 M^2 I_p^2 \tag{3.89}$$

where we assume $m_i = m(i = 1, 2, \ldots, n)$ for the sake of simplicity. As for the noise in SCM systems, the following two noise factors must be added to (3.2). They are the laser intensity noise and the *intermodulation product* (IMP) noise.

Laser intensity noise is due to the output light fluctuation of a transmitting laser. Usually, laser intensity noise is defined as *relative intensity noise* (RIN) [31,32].

$$\text{RIN} = \frac{\langle (\Delta P)^2 \rangle}{P} \tag{3.90}$$

where P is the average laser light intensity and $\langle(\Delta P)^2\rangle$ is the mean square intensity fluctuation spectral density of laser light. According to [31], RIN originates in three factors. The factors are:

1. Intrinsic (quantum) intensity fluctuation;
2. Reflection to a laser (meaning that the reflection light injects *into* a laser);
3. Spatial filtering of laser output light.

The intrinsic intensity noise is due to the statistical nature of the carrier recombination process. The reflection light from the refractive index discontinuity is reinjected into a laser, and it induces the intensity noise. The refractive index discontinuity may occur at either an optical connector or a spliced part. Intensity noise increases when only a portion of the laser output light is detected (spatial filtering). Typical values of RIN in a laser diode ranges from -130 dB/Hz to -160 dB/Hz at $I_b = 1.2 I_{\text{th}}$. RIN depends on the laser bias current [31], such as

$$\text{RIN} = C \left(\frac{I_b}{I_{\text{th}}} - 1\right)^{-3} \tag{3.91}$$

From the definition, noise due to RIN is

$$\langle i_R^2 \rangle = (\text{RIN})\, I_{\text{dc}}^2\, B = (\text{RIN})\, M^2 I_p^2\, B \tag{3.92}$$

RIN is in principle also the noise source for both baseband IM/DD and coherent schemes; however, this is neglected in Sections 3.1 and 3.2 because of the small contribution shown below. RIN is important only when I_p is large. We compare the intensity noise with shot noise. For example,

$$\frac{\langle i_R^2 \rangle}{\langle i_s^2 \rangle} = \frac{(\text{RIN})\, I_p^2\, B}{2e\, I_p\, B} = \frac{(\text{RIN})\, I_P}{2e} \tag{3.93}$$

For numerical calculations, RIN $= -135$ dB/Hz and R $= 1$ [A/W] are assumed. In the case of -40 dBm input light power, I_p is 0.1 µA and the above ratio is 0.01. This is very small. When the light power is -10 dBm, the above ratio is 10. In this case, RIN becomes important. For very-high-speed transmission higher than several gigabits per second, I_p may be large. Therefore, RIN is negligible for baseband IM/DD and coherent schemes, such as those described in Sections 3.1 and 3.2, at a bit rate lower than several gigabits per second. Here, the balanced mixer configuration, which can suppress the local laser intensity noise (RIN), is assumed in a coherent detection. When we do not use the balanced-mixer configuration, RIN of the local laser must be considered for noise calculations.

The IMP noise originates from the mixing of multicarriers by the nonlinearity of a laser. The noise due to IMP is denoted by $\langle i_{IM}^2 \rangle$. Using (3.38), (3.89), and the noises, the C/N is expressed as

$$\text{C/N} = \frac{\frac{1}{2}m^2 M^2 I_P^2}{\langle i_R^2 \rangle + \langle i_{IM}^2 \rangle + \langle i_s^2 \rangle + \langle i_d^2 \rangle + \langle i_t^2 \rangle} \quad (3.94)$$

The relationship between the C/N and the S/N depends on the modulation scheme. For example, the following equations are obtained [10]:

$$(\text{S/N}) = (\text{C/N}) \quad (\text{AM}) \quad (3.95)$$

$$(\text{S/N}) = \frac{3B(\Delta F)^2}{2f_m^3}(\text{C/N}) \quad (\text{FM}) \quad (3.96)$$

where f_m, ΔF, and B ($= 2f_m + 2\Delta F$) are the top baseband signal bandwidth, the peak frequency deviation, and the FM bandwidth, respectively. For analog TV signals (video and audio signals), a weighted signal-to-noise ratio $(S/N)_W$ is commonly used, and those for AM and FM are

$$(\text{S/N})_W \approx (\text{C/N}) \quad (\text{AM}) \quad (3.97)$$

$$(\text{S/N})_W = (\text{C/N}) + 10 \log\left[\frac{3B(\Delta F_{pp})^2}{2f_V^3}\right] + W \quad (\text{dB}) \quad (\text{FM}) \quad (3.98)$$

where ΔF_{pp} and f_v are the peak-to-peak frequency deviation and the top video baseband signal bandwidth, respectively. W is the weighting factor, and W for FM is about 14 dB (determined by the International Radio Consultive Committee (CCIR)). In this case, $B = \Delta F_{pp} + 2f_m$, where f_m is the top audio subcarrier frequency. When studio quality is required, the necessary S/N is 56 dB. In this case, the required C/N is 56 dB for AM and about 22.5 dB for FM, assuming that $B = 32$ MHz, $\Delta F_{pp} = 20$ MHz, and $f_m = 6$ MHz. As can be seen from this example, the required C/N is large in the case of AM.

3.3.2 SCM Systems Using Coherent Technology

The noise for the coherent transmission, the shot noise due to the local laser, must be taken into consideration. Since the receiver in the coherent transmission is very sensitive, RIN is less important than it is in the case of SCM systems using IM/DD. The noise is

$$\sigma^2 = \langle i_R^2 \rangle + \langle i_{IM}^2 \rangle + \langle i_s^2 \rangle + \langle i_d^2 \rangle + \langle i_t^2 \rangle + \langle i_L^2 \rangle \tag{3.99}$$

and the signal energy for one channel depends on the optical coherent modulation scheme. SCM systems using heterodyne PM is discussed in [30], and the C/N for heterodyne PM is

$$\frac{C}{N} = \frac{2 R^2 P_L P_s J_1^2(\beta) [J_0(\beta)]^{2n-2}}{\langle i_R^2 \rangle + \langle i_{IM}^2 \rangle + \langle i_s^2 \rangle + \langle i_d^2 \rangle + \langle i_t^2 \rangle + \langle i_L^2 \rangle} \tag{3.100}$$

where n and β are the number of channels and the phase modulation index, respectively. J_0 and J_1 are the Bessel functions of zero and the first order, respectively. In the derivation, the phase modulation index is assumed to be same for all n channels. The relationship between the C/N and S/N in the SCM systems using coherent transmission is the same as that discussed for the SCM systems using IM/DD.

3.4 LIMITATION DUE TO FIBER NONLINEARITY

The light intensity in a core is strong, even for relatively modest laser power, because of a small core diameter. When used for optical transmission, fiber length is generally long, and this results in a long interaction length. Both strong light intensity and long interaction length cause several nonlinear effects in silica fibers [33–35]. Typical effects are *stimulated Raman scattering* (SRS), *stimulated Brillouin scattering* (SBS), *self-phase modulation* (SPM), *carrier-induced phase modulation* (CIP), and *four-wave mixing* (FWM). These influence the transmission quality and impose limitations on systems: the noise increase due to crosstalk or phase noise and the limitation of maximum input power into a fiber. Although nonlinear effects are harmful to ordinary transmission systems, several useful applications using the fiber nonlinear effects have been investigated and used. Degradation and application in relation to nonlinear effects are listed in Table 3.5. They are explained below.

3.4.1 SRS

Raman scattering originates in the interaction of light and molecular vibration in a fiber. This effect can be viewed as the modulation of light by molecular vibrations, and it creates lights in the upper and lower bands. These frequency-shifted lights are called *anti-Stokes light* and *Stokes light*. The Raman effect not only causes the frequency shift of the incident light, but also optical gain or optical absorption at the

Table 3.5
Fiber Nonlinearity

Phenomenon	Degradation of Transmission Sys.	Application
SRS	Crosstalk in WDM Crosstalk in OFDM	Raman amplifier Fiber Raman laser
SBS	Crosstalk in WDM Crosstalk in OFDM Limitation to maximum input power	Brillouin amplifier Brillouin OTDA
SPM	Phase noise in PSK or PM	Optical soliton
CIP	Cross phase modulation in PSK or PM	
FWM	Crosstalk in OFDM	

shifted frequency. The Stokes light is amplified by a strong pump light, while the anti-Stokes light is absorbed. The length dependence of these light powers is expressed as

$$P_s(L) = P_s(0) \exp(gI_0L_e) \tag{3.101}$$

$$P_a(L) = P_a(0) \exp(-gI_0L_e) \tag{3.102}$$

where P_s and P_a are the Stokes and anti-Stokes light powers. I_0, L, L_e, and g are the pump light intensity, the fiber length, the effective fiber length, and the Raman gain, respectively. The effective length L_e is introduced by considering the attenuation of pump light [33,34]. The derivation of L_e is shown below. The intensity of pump light is

$$I(x) = I_0 \exp(-\alpha x) \tag{3.103}$$

where α is the fiber loss, including splice loss, and x is the fiber length. The Stokes light is amplified, and the power at $x = x + \Delta x$ is

$$P_s(x + \Delta x) = P_s(x) \exp[gI(x)\Delta x] \tag{3.104}$$

Then the following equation holds:

$$\frac{dP_s}{dx} = P_s gI_0 \exp(-\alpha x) \tag{3.105}$$

Integration of this equation results in

$$\ln\left[\frac{P_s(x)}{P_s(0)}\right] = gI_0 \frac{1 - \exp(-\alpha x)}{\alpha} \quad (3.106)$$

Then L_e is expressed as

$$L_e \equiv \frac{1 - \exp(-\alpha L)}{\alpha} \quad (3.107)$$

$L_e = 1/\alpha$ for $\alpha L \gg 1$, and $L_e = L$ for $\alpha L \ll 1$. The intensity of pump light I_0 is expressed as $I_0 = P/A_e$, where P is the light power and A_e is the effective area. A_e is related to the fiber core and is determined by the overlap of pump and signal light fields. In a single-mode fiber, A_e nearly equals $\pi\omega^2$ (ω is the mode field radius). The polarization states for the pump light and the Stokes or anti-Stokes light influence the efficiency of SRS. Equations (3.101) and (3.102) are for the same polarization states. Raman amplification is observed even for input lights with orthogonal polarization states, and the measured gain in an ordinary single-mode fiber is reported to be nearly half of that for parallel polarization states [36]. In general, two equations are written using the parameter s:

$$P_s(L) = P_s(0) \exp[gP_p s L_e / A_e] \quad (3.108)$$

$$P_a(L) = P_a(0) \exp[-gP_p s L_e / A_e] \quad (3.109)$$

where P_p is the pump light power and s represents the polarization state. $s = 1$ for polarization maintenance fibers and $s = 0.5$ for ordinary fibers. The gain exists for lights traveling in the same direction as well as the opposite direction with respect to the pump light. According to [33], the critical power P_c is expressed as

$$P_c = \frac{16 A_e}{g s L_e} = \frac{16 A_e \alpha}{g s [1 - \exp(-\alpha L)]} \quad (3.110)$$

and this critical power P_c is defined as the power in which the Stokes light power equals the pump light power. Therefore, half of the input power converts to the Stokes light power when the input power is P_c. Typical values for a silica single-mode fiber are:

Gain bandwidth of SRS: 10,000 GHz;

Gain peak value of SRS: 7×10^{-12}–2×10^{-11} cm/W;

Frequency shift of SRS: 0–15,000 GHz (about 0–120 nm);

Critical power P_c of SRS: 1W–5W.

The phenomena of SRS for single and multichannel transmission are shown in Figure 3.21. In the case of a single channel, the frequency-shifted lightwaves propagate in both directions. The backward light may influence the laser diode lasing conditions and cause degradation in some systems. However, the intensity of SRS light is not so strong when compared to SBS light intensity for the same input light power. Therefore, SRS in the case of a single channel is not a problem. In the case of multichannel transmission, there exists the interaction between the channels [37]. The lower frequency (longer wavelength λ_2) light is amplified by the higher frequency (shorter wavelength λ_1) light. In this case, the intensity of pump light (higher frequency light) decreases. These cause the crosstalk. According to [38], the following result for n-channel systems with 0.5-dB penalty is obtained:

$$nP(n-1)\,\Delta f < 330 \text{ GHz W} \tag{3.111}$$

where Δf and P are the channel spacing and the power per channel, respectively. The triangular gain profile with bandwidth 15,000 GHz is assumed in obtaining this equation. This equation indicates that the degradation due to SRS is acceptable when the product of total power and total optical bandwidth is smaller than 330 GHz W.

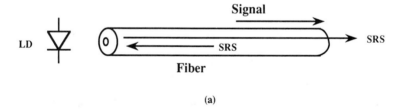

(a)

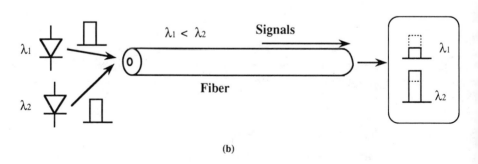

(b)

Figure 3.21 Stimulated Raman scattering: (a) single-channel transmission; (b) multichannel transmission.

Since the gain bandwidth of SRS in a fiber is very broad, this effect must be taken into consideration for designing WDM systems as well as OFDM systems.

It has been indicated that the group velocity dispersion influences the SBS effect, and the allowable power P in (3.111) is 0 to 3 dB higher for lightwaves with nonequal group velocity [39]. The SRS applications are the Raman amplifier and the fiber Raman laser [40,41]. The fiber Raman laser can be widely tuned over about 50 nm because of its wide gain bandwidth.

3.4.2 SBS

The Brillouin scattering effect originates in the interaction of the light and acoustic vibrations in a fiber. SBS is similar to SRS except for the following differences. The Brillouin gain is about two orders of magnitude greater than the Raman gain. The Brillouin frequency shift and the gain bandwidth are much smaller. The comparison of SRS and SBS is shown schematically in Figure 3.22. Typical values for a silica single-mode fiber are:

Gain bandwidth of SBS: 20 to 100 MHz;
Gain peak value of SBS: 4×10^{-9} cm/W;
Frequency shift of SBS: 10 to 13 GHz.

Contrary to SRS, SBS exists only for the lightwave propagating in the opposite direction from the pump light. The Brillouin gain bandwidth Δf_B is narrow (Δf_B = 20 to 100 MHz); therefore, the effective gain decreases approximately by $\Delta f_B/\Delta f_s$ for sufficiently large signal light linewidth Δf_s ($\Delta f_s \gg \Delta f_B$). Ordinary laser diodes have broader linewidth than Δf_B. In this case, the power at $x = -L$ is

$$P(-L) = P(0) \exp\left[g \frac{\Delta f_B}{\Delta f_s} P_p s L_e / A_e \right] \tag{3.112}$$

The critical power P_c of SRS for light with linewidth Δf_s is

$$P_c = \frac{21 A_e}{g s L_e} \left(\frac{\Delta f_s}{\Delta f_B}\right) = \frac{21 A_e \alpha}{g s [1 - \exp(-\alpha L)]} \left(\frac{\Delta f_s}{\Delta f_B}\right) \tag{3.113}$$

For some lasers used for coherent transmission, linewidth Δf_s is comparable or less than Δf_B. In this case, the ratio $\Delta f_s/\Delta f_B$ in (3.112) and (3.113) equals 1, and the SBS effect is more important.

Numerical examples for P_c

Example 1:
Fiber parameter: $\omega = 4$ μm, $L = 30$ km, $\alpha = 0.2$ dB/km, $s = 0.5$

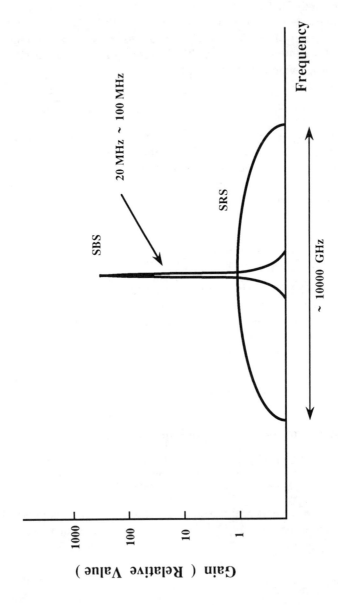

Figure 3.22 Comparison of SRS and SBS.

SBS parameter: $g = 4 \times 10^{-9}$ cm/w, $\Delta f_B = 100$ MHz
Linewidth of LD: $\Delta f_s < 100$ MHz
Critical power P_c of SBS is about 3 mW.

Example 2:

Fiber parameter: $\omega = 4$ μm, $L = 5$ km, $\alpha = 0.5$ dB/km, $s = 0.5$
SBS parameter: $g = 4 \times 10^{-9}$ cm/w, $\Delta f_B = 100$ MHz
Linewidth of LD: $\Delta f_s = 1$ GHz
Critical power P_c of SBS is about 140 mW.

Example 1 is assumed for trunk transmission systems using a narrow LD linewidth, and the critical power is comparable with LD output power. On the other hand, example 2 is assumed for subscriber loop systems using a relatively broad LD linewidth, and this example shows that SBS is less important for systems with these parameters.

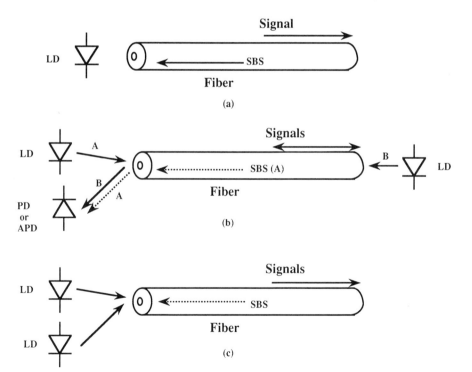

Figure 3.23 Stimulated Brillouin scattering: (a) unidirectional transmission; (b) bidirectional transmission; (c) multichannel transmission.

The influence of SBS on optical transmission systems is shown in Figure 3.23. In the case of a unidirectional transmission with a single channel, the strong backward light returns to a laser when the input light nearly has the critical power. In some systems, without using isolators, the backward light may be harmful. This also imposes the input power limitation and most of the input power cannot transmit to the fiber output for sufficiently large input power [42]. In the case of bidirectional transmission systems, the backward light may cause the crosstalk, as shown in Figure 3.23(b) [43]. In a TCM system, there is no problem. For multichannel transmission systems like those shown in Figure 3.23(c), each channel interacts with a fiber independently because of a narrow bandwidth Δf_B and small frequency shift. That is, the critical power P_c does not depend on the number of channels.

The critical power P_c depends on the modulation scheme and is slightly different from the value for CW light. In the case of ASK, FSK, and PSK coherent transmission using a laser with a narrow linewidth Δf_s ($\Delta f_s < \Delta f_B$), the dependence of P_c on the transmission schemes was investigated in [44]. One example of experimentally obtained results is that P_c in PSK at 400 Mb/s with a pseudorandom pattern is nearly double the CW P_c value. The CW P_c value is about 15 mW for $\alpha = 0.4$ dB/km and $L = 10$ km. In the experiment, a light from a laser with $\Delta f_s < 20$ MHz was PSK-modulated by using an external $LiNbO_3$ phase modulator. For the application of SBS, the Brillouin amplifier [45] and Brillouin Optical-Fiber Time Domain Analysis (OTDAs) [46] have been investigated. The Brillouin shift depends on the velocity of the acoustic wave, which is influenced by the strain in a fiber. Therefore, the Brillouin OTDA, which measures Brillouin shift, can measure the strain in a fiber.

3.4.3 Intensity-Dependent Refractive Index

The nonlinear effect in a fiber due to the intensity-dependent refractive index is expressed as

$$n = n_0 + n_2 I \tag{3.114}$$

where n_0 and n_2 are the ordinary and intensity-dependent refractive indices. The phase of transmitted light through a fiber with length L is

$$\phi = \frac{2\pi n L}{\lambda} \tag{3.115}$$

Therefore, the AM (IM)-PM conversion occurs. In the case of a single channel, as shown in Figure 3.24(a), SPM occurs and the intensity-dependent phase $\Delta\phi$ may cause the degradation in the PM or PSK system. If the amplitude of PSK-modulated

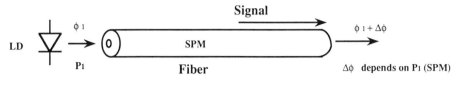

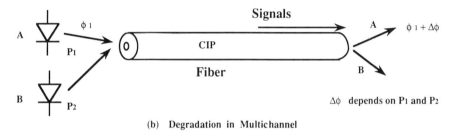

Figure 3.24 SPM and CIP: (a) degradation in single-channel transmission; (b) degradation in multichannel transmission.

light is ideally constant, there is no $\Delta\phi$ (no degradation). In the case of multichannel transmission, as shown in Figure 3.24(b), the intensity-dependent phase $\Delta\phi$ for one channel (e.g., channel A) depends on the summing up of light intensity of all channels. This CIP may degrade the PM or PSK system performance when residual AM (IM) components exist for phase-modulated light. It is anticipated that crosstalk in PM or PSK signal light is large when used with AM or IM signal light. For example, one channel uses PSK and $(n-1)$ channels use IM in n-channel transmission systems. In this case, CIP due to the total optical power of n channels must be taken into consideration in the design of the PSK system.

Optical soliton transmission is the typical application of the intensity-dependent refractive index effect. A soliton is constructed by the SPM effect and the *group velocity dispersion* (GVD) effect in a fiber [47,48]. A soliton pulse can propagate without its shape being changed by the cancellation of pulse broadening and pulse compression based on two effects. Optical soliton transmission has been investigated mainly for high-speed trunk systems.

3.4.4 FWM

Mixing of two or more lightwaves at different wavelengths in a fiber produces new lightwaves at other wavelengths [33]. This effect is FWM. FWM is analogous to the IMP effect in electric waves, described in Section 3.3. FWM is very important when designing multichannel transmission systems, especially for OFDM systems, because FWM causes the crosstalk. The effect of FWM is shown schematically in Figure 3.25: three lightwaves from LDs are injected into a single-mode fiber. Nine new lightwaves are generated by FWM. In the FWM process, four photons are involved in the process, and their frequencies and propagation constants are denoted by f_i, f_j, f_k, f_{ijk} and β_i, β_j, β_k, β_{ijk}. Since the photon energy is hf (h is the Plank constant), the energy conservation requires

$$f_{i,j,k} = f_i + f_j - f_k \tag{3.116}$$

and the conservation of momentum requires

$$\beta_{i,j,k} = \beta_i + \beta_j - \beta_k \tag{3.117}$$

Equation (3.117) is known as phase matching and is derived by the fact that the momentum is proportional to $h\beta$. In the case of the phase mismatch $\Delta\beta$, the efficiency of FWM decreases. The phase mismatch $\Delta\beta$ is

$$\Delta\beta = \beta_{i,j,k} - (\beta_i + \beta_j - \beta_k) \tag{3.118}$$

The power of the generated light for fiber length L is [49]

$$P_{i,j,k}(L) = K\left[\frac{P_i P_j P_k}{A_e^2}\right]\left[\left|\frac{\exp(j\Delta\beta - \alpha)L - 1}{j\Delta\beta - \alpha}\right|^2\right]\exp(-\alpha L) \tag{3.119}$$

where K is constant and P_i, P_j, and P_k are the powers for lightwaves with f_i, f_j, and f_k, respectively. A_e is the effective area and α is the fiber loss. From (3.119), the

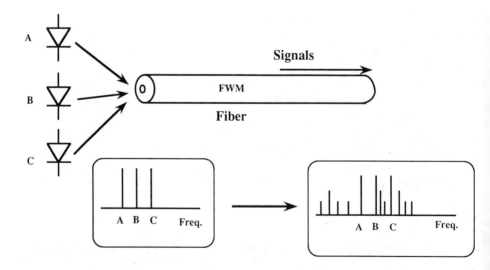

Figure 3.25 Four-wave mixing.

efficiency depends on the fiber length L, the fiber loss α, the fiber mode field diameter, the powers, and the phase mismatch $\Delta\beta$. It is pointed out that both channel frequency separation Δf and the fiber chromatic dispersion influence the phase mismatch $\Delta\beta$. These factors cause the lightwaves to have different group velocities, and they destroy the phase matching. The wider channel frequency separation decreases the FWM efficiency, and negligible efficiency is expected when $\Delta f > 30$ GHz for ordinary single-mode fibers and $\Delta f > 100$ GHz for dispersion-shifted fibers at wavelength $\lambda = 1.55$ μm [50].

REFERENCES

[1] Geels, R. S., and L. A. Coldren, "Low Threshold, High Power, Vertical-Cavity Surface-Emitting Lasers," *Electron. Lett.*, Vol. 27, 1991, p. 1984.
[2] Yokoyama, H., and S. D. Brorson, "Rate Equation Analysis of Microcavity Lasers," *J. Appl. Phys.*, Vol. 66, 1989, p. 4801.
[3] Personick, S. D., "Receiver Design for Digital Fiber Optic Communication Systems, I," *B.S.T.J.*, Vol. 52, 1973, p. 843.
[4] Personick, S. D., "Receiver Design for Digital Fiber Optic Communication Systems, II," *B.S.T.J.*, Vol. 52, 1973, p. 875.
[5] Smith, R. G., and S. D. Personick, "Receiver Design for Optical Fiber Communication System" Chap. 4 in *Semiconductor Devices for Optical Communications*, H. Kressel, ed., Springer Verlag, 1980.
[6] Gagliardi, R. M., and S. Karp, *Optical Communications*, John Wiley & Sons.
[7] Personick, S. D., P. Balan, J. H. Bobsin, and P. R. Kumar, "A Detailed Comparison of Four Approaches to the Calculation of the Sensitivity of Optical Fiber System Receivers," *IEEE Trans. Commun.*, Vol. COM-25, 1977, p. 541.
[8] Muoi, T. V., "Receiver Design for High-Speed Optical-Fiber Systems," *IEEE J. of Lightwave Technol.*, Vol. LT-2, 1984, p. 243.
[9] Saito, T., Y. Sunohara, K. Fukagai, S. Ishikawa, N. Hennmi, S. Fujita, and Y. Aoki, "High Receiver Sensitivity at 10 Gb/s Using an Er-Doped Fiber Preamplifier Pumped With a 0.98 μm Laser-Diode," *Conf. on Optical Fiber Communication '91 (OFC'91)*, San Diego, PD14, 1991.
[10] Stein, S., and J. J. Jones, *Modern Communication Principles*, McGraw-Hill, 1965.
[11] Yamamoto, Y., "Receiver Performance Evaluation of Various Digital Optical Modulation-Demodulation Systems in the 0.5 μm–10 μm Wavelength Region," *IEEE J. of Quantum Electron.*, Vol. QE-16, 1980, p. 1251.
[12] Yamamoto, Y., and T. Kimura, "Coherent Optical Fiber Transmission Systems," *IEEE J. of Quantum Electron.*, Vol. QE-17, 1981, p. 919.
[13] Okshi, T., K. Emura, K. Kikuchi, and R. T. Kersten, "Computation of Bit-Error Rate of Various Heterodyne and Coherent-Type Optical Communication Schemes," *J. Opt. Commun.*, Vol. 2, 1981, p. 89.
[14] Nosu, K., "Advanced Coherent Lightwave Technologies," *IEEE Communications Magazine*, Vol. 26, February 1988, p. 15.
[15] Takachio, N., and K. Iwashita, "Compensation of Fibre Chromatic Dispersion in Optical Heterodyne Detection," *Electron. Lett.*, Vol. 24, 1988, p. 108.
[16] Smith, D. W., "Technique for Multigigabit Coherent Optical Transmission," *IEEE J. of Lightwave Technol.*, Vol. LT-5, 1987, p. 1466.
[17] Abbas, G. L., V. W. S. Chan, and T. K. Yee, "Local-Oscillator Excess-Noise Suppression for Homodyne and Heterodyne Detection," *Opt. Lett.*, Vol. 8, 1983, p. 419.

[18] Okoshi, T., "Ultimate Performance of Heterodyne/Coherent Optical Fiber Communications," *IEEE J. of Lightwave Technol.*, Vol. LT-4, 1986, p. 243.
[19] Hodgkinson, T. G., R. A. Harmon, and D. W. Smith, "Polarization-Insensitive Heterodyne Detection Using Polarization Scrambling," *Electron. Lett.*, Vol. 23, 1987, p. 513.
[20] Salz, J., "Modulation and Detection for Coherent Lightwave Communications," *IEEE Communications Magazine*, Vol. 24, June 1986, p. 38.
[21] Linke, R. A., and A. H. Gnauck, "High-Capacity Coherent Lightwave Systems," *IEEE J. of Lightwave Technol.*, Vol. 6, 1988, p. 1750.
[22] Chikama, T., S. Watanabe, T. Naito, H. Onaka, T. Kiyonaga, Y. Onoda, H. Miyata, M. Suyama, M. Seino, and H. Kuwahara, "Modulation and Demodulation Techniques in Optical Heterodyne PSK Transmission Systems," *IEEE J. of Lightwave Technol.*, Vol. 8, 1990, p. 309.
[23] DeLange, O. E., "Wideband Optical Communication Systems: Part II—Frequency Division Multiplexing," *Proc. IEEE*, Vol. 58, 1970, p. 1683.
[24] Bachus, E.-J., R. P. Braun, W. Eutin, E. Großmann, H. Foisel, K. Heimes, and B. Strebel, "Coherent Optical Fiber Subscriber Line," *Proc. IOOC—ECOC'85*, Vol. III, 1985, p. 61.
[25] Nosu, K., H. Toba, and K. Iwashita, "Optical FDM Transmission Technique," *IEEE J. of Lightwave Technol.*, Vol. LT-5, 1987, p. 1301.
[26] Kazovsky, L. G., "Multichannel Coherent Optical Communications Systems," *IEEE J. of Lightwave Technol.*, Vol. LT-5, 1987, p. 1095.
[27] Darcie, T. E., "Subcarrier Multiplexing for Multiple-Access Lightwave Networks," *IEEE J. of Lightwave Technol.*, Vol. LT-5, 1987, p. 1103.
[28] Olshansky, R., V. A. Lanzisera, and P. M. Hill, "Subcarrier Multiplexed Lightwave Systems for Broad-Band Distribution," *IEEE J. of Lightwave Technol.*, Vol. 7, 1989, p. 1329.
[29] Way, W. I., "Subcarrier Multiplexed Lightwave System Design Consideration for Subscriber Loop Applications," *IEEE J. of Lightwave Technol.*, Vol. 7, 1989, p. 1806.
[30] Gross, R., and R. Olshansky, "Multichannel Coherent FSK Experiments Using Subcarrier Multiplexing Techniques," *IEEE J. of Lightwave Technol.*, Vol. 8, 1990, p. 406.
[31] Sato, K., "Intensity Noise of Semiconductor Laser Diodes in Fiber Optic Analog Video Transmission," *IEEE J. of Quantum Electron.*, Vol. QE-19, 1983, p. 1380.
[32] Kim, H. B., R. Maciejko, and J. Conradi, "Effect of Laser Noise on Analogue Fibre Optic Systems," *Electron. Lett.*, Vol. 16, 1980, p. 919.
[33] Stolen, R. H., "Nonlinear Properties of Optical Fibers," in *Optical Fiber Telecommunications*, S. E. Miller and A. G. Chynoweth, eds., Academic Press, 1979.
[34] Stolen, R. H., "Nonlinearity in Fiber Transmission," *Proc. IEEE*, Vol. 68, 1980, p. 1232.
[35] Chraplyvy, A. R., "Limitations on Lightwave Communications Imposed by Optical-Fiber Nonlinearities "*IEEE J. of Lightwave Technol.*, Vol. 8, 1990, p. 1548.
[36] Ikeda, M., "Stimulated Raman Amplification Characteristics in Long Span Single-Mode Silica Fibers," *Optics. Comm.*, Vol. 39, 1981, p. 148.
[37] Chraplyvy, A. R., and P. S. Henry, "Performance Degradation Due to Stimulated Raman Scattering in Wavelength-Division-Multiplexed Optical-Fiber Systems," *Electron. Lett.*, Vol. 19, 1983, p. 641.
[38] Chraplyvy, A. R., "Optical Power Limits in Multichannel Wavelength-Division-Multiplexed System to Stimulated Raman Scattering," *Electron. Lett.*, Vol. 20, 1984, p. 58.
[39] Cotter, D., and A. M. Hill, "Stimulated Raman Crosstalk in Optical Transmission: Effect of Group Velocity Dispersion," *Electron. Lett.*, Vol. 20, 1984, p. 85.
[40] Ippen, E. P., "Low Power Quasi-CW Raman Oscillator," *Appl. Phys. Lett.*, Vol. 16, 1970, p. 303.
[41] Stolen, R. H., E. P. Ippen, and A. R. Tynes, "Raman Oscillation in Glass Optical Waveguide," *Appl. Phys. Lett.*, Vol. 20, 1972, p. 62.
[42] Uesugi, N., M. Ikeda, and Y. Sasaki, "Maximum Single-Frequency Input Power in a Long Optical Fiber Determined by Stimulated Brillouin Scattering," *Electron. Lett.*, Vol. 17, 1981, p. 379.

[43] Waarts, R. G., and R. P. Braun, "Crosstalk Due to Stimulated Brillouin Scattering in Monomode Fibre," *Electron. Lett.*, Vol. 21, 1985, p. 1114.
[44] Aoki, Y., K. Tajima, and I. Mito," Input Power Limits of Single-Mode Optical Fibers Due to Stimulated Brillouin Scattering in Optical Communication Systems," *IEEE J. of Lightwave Technol.*, Vol. 6, 1988, p. 710.
[45] Atkins, C. G., D. Coter, D. W. Smith, and R. Wyatt, "Application of Brillouin Amplification in Coherent Optical Transmission," *Electron. Lett.*, Vol. 22, 1986, p. 556.
[46] Horiguchi, T., and M. Tateda, "BOTDA-Nondestructive Measurement of Single-Mode Optical Fiber Attenuation Characteristics Using Brillouin Interaction: Theory," *IEEE J. of Lightwave Technol.*, Vol. 7, 1989, p. 1170.
[47] Hasegawa, A., and F. Tappert, "Transmission of Stationary Nonlinear Optical Pulses in Dispersive Dielectric Fibers I, Anomalous Dispersion," *Appl. Phys. Lett.*, Vol. 23, 1973, p. 142.
[48] Hasegawa, A., and F. Tappert, "Transmission of Stationary Nonlinear Optical Pulses in Dispersive Dielectric Fibers II, Normal Dispersion," *Appl. Phys. Lett.*, Vol. 23, 1973, p. 171.
[49] Hill, K. O., D. C. Johnson, B. S. Kawasaki, and R. I. MacDonald, "CW Three-Wave Mixing in Single-Mode Optical Fibers," *J. Appl. Phys.*, Vol. 49, 1978, p. 5098.
[50] Shibata, N., R. P. Braun, and R. G. Waarts, "Phase-Mismatch Dependence of Efficiency of Wave Generation Through Four-Wave Mixing in a Single-Mode Optical Fiber," *IEEE J. of Quantum Electron.*, Vol. QE-23, 1987, p. 1205.

Chapter 4
Optical Receivers

The performance of an optical receiver determines the transmission system performance after the transmission schemes and components are determined. Therefore, designing a good and inexpensive receiver is very important. In this chapter, receiver designs are classified, explained, and compared.

4.1 CLASSIFICATION AND REQUIREMENTS OF RECEIVER DESIGN

There are several optical receiver designs, which have been developed to meet system requirements, such as high receiver sensitivity, wide dynamic range, bit rate transparency, bit pattern independence, and fast acquisition time. In some systems, wavelength independence may be required. The typical optical receiver configuration for digital baseband transmission is shown in Figure 4.1. Detected light signals are amplified by a preamplifier and a main amplifier. An *equalizer* (EQL) is used when needed. An *automatic gain control* (AGC) circuit is used to realize a nearly constant amplitude signal level for various optical input power levels. A *filter* (FIL) is used for noise reduction, and the filtered signals input the decision circuit and then the original data signals are recovered. For decision circuits, timing signals must be provided, and they are usually extracted from the transmitted signals. The configuration of optical receivers is similar to that of the conventional PCM receivers except for the front-end circuit. The configuration of optical receivers used for heterodyne detection in optical coherent transmission systems or for SCM transmission systems is slightly different from that for digital baseband transmission. This configuration is shown in Figure 4.2. After the conversion from optical signals and amplification, detected signals input an AGC circuit and a BPF. The band-pass filtered signals are demodulated by the appropriate demodulation (demod.) method, which is explained in the previous chapter. Analog signals are taken out with an LPF. The digital signals are recovered with a decision circuit. The front-end circuit in Figure 4.2 is not required to detect baseband signals. Therefore, different ap-

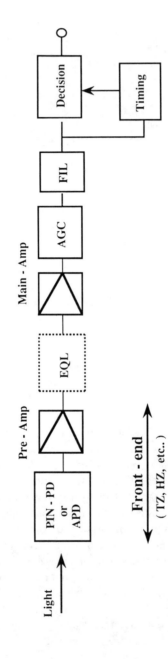

Figure 4.1 Optical receiver for baseband digital transmission.

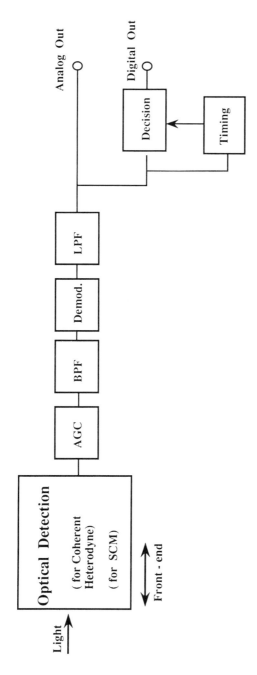

Figure 4.2 Optical receiver for coherent heterodyne or SCM transmission.

proaches are possible for designing these front-end circuits. From the viewpoints of receiver sensitivity and dynamic range, front-end is the most important design issue. There are *high-impedance design* (HZ design), *transimpedance design* (TZ design), *low-impedance design* (LZ design), resonance-type design, and peaking design. TZ design is used for practical transmission systems for baseband digital transmission systems. Resonance-type design is intended for coherent transmission systems, and peaking design is for high-speed transmission systems. These designs are explained later.

The requirements for optical receivers are different for application areas. In subscriber loop systems, the locations of subscribers are scattered from the nearest to the farthest location in the same subscriber area. Therefore, the transmission length L between a TO and a subscriber is not the same for all subscribers, and the received power P differs. The same receiver must be used in the same subscriber area (if possible, all of the existing subscriber area), because a subscriber might move to a new location. This requires a wide dynamic range for optical receivers. When we use a PON, receiver sensitivity must be high to compensate for a branch loss of a star coupler. In the case of TCM, optical burst signals are used and fast acquisition time is required for optical receivers. These requirements are shown in Figure 4.3.

In digital transmission, especially for data transmission, bit pattern independence is desirable, which means that a receiver has the same receiving performance for any bit pattern. In most cases, long strings of zeros or ones can be avoided at a sending side (a transmitter) by using a scrambling technique in which data are scrambled to obtain a quasirandom pattern at a transmitter, and the scrambled data are transmitted through a fiber and recovered by descrambling at a receiver. In the worst case, the scrambled result happens to have long strings of zeros or ones for a special data pattern. When NRZ code is used, the long strings of zeros or ones indicate the absence of transitions, and this causes difficulties in the timing extraction. An appropriate line code with many transitions must be chosen for some systems. For example, we can use CMI code instead of NRZ code if the system has sufficient bandwidth. The long period of no transition also causes the dc reference level to wander when an ac coupling is used in the receiver. To prevent this, a time constant for the coupling must be designed to be sufficiently long for avoiding a complete discharge of the capacitor in a no-transition period. It requires a large coupling capacitor in general, and it causes difficulties when integrated into an IC or LSI. One countermeasure to this low-frequency cutoff problem is the use of a quantized feedback circuit. Another method to ensure the dc reference level is the use of a dc coupling design. In this design, drifts and offsets in amplifiers caused by temperature change or heat accumulation must be carefully avoided.

Some systems may require bit rate transparency and wavelength and polarization independence. Bit rate transparency is usually not required for telecommunication systems, in which the bit rate is defined in digital hierarchy. According to [1], the bit rates are normally set by the users in data buses and LANs. In such applications, it is desirable for optical receivers to operate over a wide range of bit

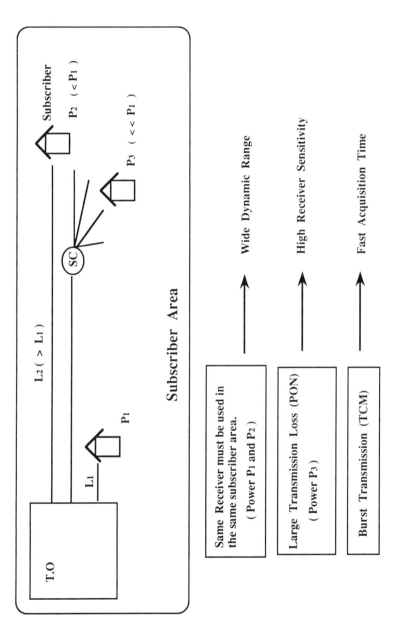

Figure 4.3 Requirements for receivers in subscriber loop systems.

rates with no or slight modifications. Wavelength independence also is not required explicitly for telecommunication systems, in which the wavelength used is determined and not changed ordinarily. By using an InGaAs APD or an InGaAs pin-PD, wavelength independence for a longer wavelength band (such as 1.3 and 1.5 μm) is easily realized. Polarization independence is required because almost all systems currently use ordinary single-mode fibers instead of polarization maintaining fibers.

4.2 SEVERAL FRONT-END DESIGNS

There, the HZ, TZ, LZ, resonance-type, and peaking designs are explained with regard to designing a front-end.

4.2.1 High-Impedance Design

In HZ design, a pin-PD or an APD is terminated by high impedance (resistance R_L) [2,3]. The thermal noise from resistance R_L is

$$\langle i_t^2 \rangle = \frac{4kT}{R_L} B \qquad (4.1)$$

where, k, T, and R_L are the Boltzmann constant, the temperature, and the load resistance, respectively. In this equation, the waveform effect of pulse is ignored for simplicity. High sensitivity is expected by lowering this thermal noise, which indicates that R_L must have high impedance. For ordinary applications, the bandwidth of a single-mode fiber is large enough. Therefore, the bandwidth is limited by an electric circuit and, in this case, a front-end circuit. In HZ design, resistance R_L and capacitance C_T are located in parallel, and its transfer function $Z(f)$ is

$$Z(f) = \frac{V}{I} = \frac{A\,R_L}{1 + j2\pi f R_L C_T} \qquad (4.2)$$

and the cutoff frequency is

$$f_c = \frac{1}{2\pi R_L C_T} \qquad (4.3)$$

where C_T is the total capacitor, which includes a capacitor C_t of a pin-PD or an APD, an input capacitor C_a of a preamplifier, and a stray capacitor C_s. A in (4.2) is an amplification gain in a preamplifier and is treated as a constant in this equation. As R_L increases, f_c becomes narrow and an EQL is required. The basic configuration with an EQL is shown in Figure 4.4. The function of the EQL is to compensate the narrowed bandwidth caused by a high-impedance termination. The basic function of the EQL is explained by Figure 4.5. The A and B points in this figure correspond to the points indicated in Figure 4.4. As an explanation, a simple *resistance-capacitance* (RC) circuit in [3] as an EQL is shown in Figure 4.6(a). A transfer function of this RC-EQL circuit has a zero at

$$f_1 = \frac{1}{2\pi R_1 C_1} \qquad (4.4)$$

and a pole at

$$f_2 = \frac{R_1 + R_2}{2\pi R_1 R_2 C_1} \qquad (4.5)$$

By matching the cutoff frequency f_c at the A point and the cutoff frequency f_1 of the EQL transfer function (i.e., $f_1 = f_c$), a flat frequency response up to the frequency f_2 is obtained at the B point. To obtain a good equalization performance, a precise match is required. No equalization or poor equalization causes intersymbol interference, which is the overlapping of adjacent signal pulses (symbols). Intersymbol interference degrades the receiver performance. There are many circuits for the EQL using several types of filters, such as a simple RC filter, a Tomson filter, a transversal filter (tapped-delay line filter), and a Kalman filter. One example among these is shown in Figure 4.6(b).

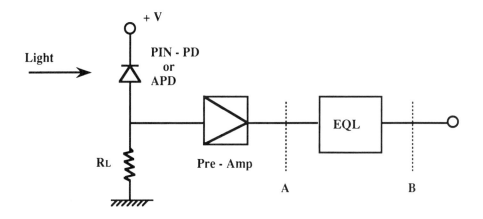

Figure 4.4 HZ design.

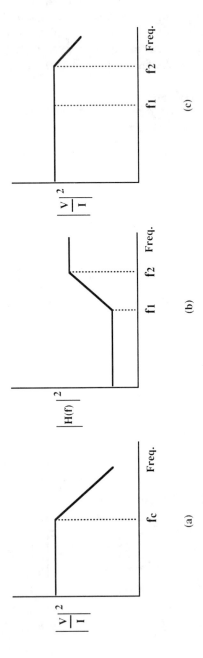

Figure 4.5 EQL in HZ design: (a) output of preamp (at A point); (b) EQL; (c) output of EQL (at B point).

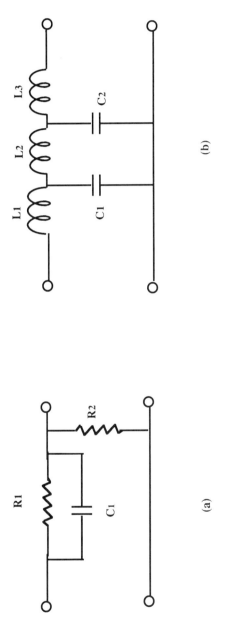

Figure 4.6 Examples of EQL: (a) RC filter EQL; (b) Tomson filter EQL.

In HZ design, the highest receiver sensitivity is realized among HZ, TZ, and LZ designs because the minimum thermal noise is obtained by the high impedance. However, the dynamic range of the receiver is narrow, which will be discussed later.

4.2.2 Transimpedance Design

The basic TZ design is shown in Figure 4.7, and this design uses the feedback with a feedback resistance R_f [4,5]. TZ designs are classified according to the types of feedback methods (shown in Figure 4.8). Three types of TZ design are listed in this figure: ordinary TZ design, active feedback design, and optical feedback design. In ordinary TZ design, a feedback is made by an ordinary resistance R_f, and this feedback method is usually adopted. The active feedback design uses an active device as R_f instead of ordinary resistance [6]. In the optical feedback design, a light path is used as a feedback loop path instead of an electrical loop path [7–9].

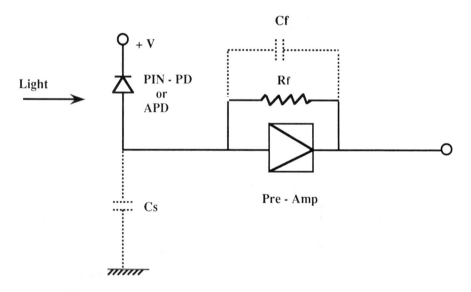

Figure 4.7 TZ design.

Ordinary TZ Design

In ordinary TZ design, the feedback is caused by ordinary feedback resistance R_f, as shown in Figure 4.7. The preamplifier in this figure is designed to have high input

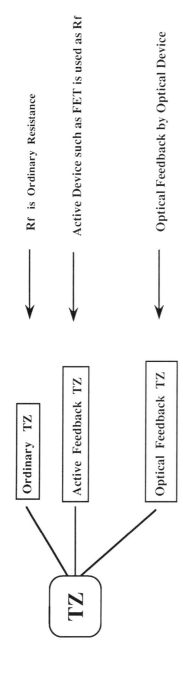

Figure 4.8 Classification of TZ design.

impedance. With sufficiently large input impedance of the preamplifier, the transfer function $Z(f)$ in TZ design is

$$Z(f) = \frac{R_f}{1 + j2\pi f R_f (C_f + C_T/A)} \tag{4.6}$$

and the cutoff frequency is

$$f_c = \frac{A}{2\pi R_f (AC_f + C_T)} \tag{4.7}$$

where C_T is the total capacitor, which includes a capacitor C_t of a pin-PD or an APD, an input capacitor C_a, and a stray capacitor C_s. An actual resistance is not an ideal resistance, and it has small capacitance value; this effect is included as a capacitor C_f. A is an amplification gain in a preamplifier. When A takes a large value, then the cutoff frequency in (4.7) tends to be $1/(2\pi f R_f C_f)$. Since the value of C_f is much smaller than the value of C_T, the bandwidth becomes large.

The thermal noise from the resistance R_f is

$$\langle i_t^2 \rangle = \frac{4kT}{R_f} B \tag{4.8}$$

and higher sensitivity is expected by designing R_f as a large value. However, the cutoff frequency decreases, as indicated by $1/(2\pi f R_f C_f)$. In TZ design, R_f is designed to be a smaller value than the value of resistance R_L in HZ design. Therefore, the sensitivity of a receiver designed by TZ design is lower than that designed by HZ design. However, wider dynamic range, which will be discussed later, and wider bandwidth are realized in a TZ-designed receiver.

Active Feedback Design

Feedback design in which an active device is used as R_f instead of an ordinary resistance was proposed and named as active feedback design in [6]. In TZ design, higher sensitivity is expected by using large-value R_f. Ordinary large-value feedback resistors have a relatively large value parasitic shunt capacitance C_f. C_f is about 0.1 pF for conventional large-value resistors [6]. With large C_f and large R_f, the TZ design is similar to HZ design and the merits of TZ design decrease. To avoid the increase of shunt capacitance C_f, an active feedback design is proposed. According to [6], two approaches for a pin-PD were taken in active feedback design. They are shown in Figure 4.9. In approach 1, a micro-FET in an IC is used. The FET acts

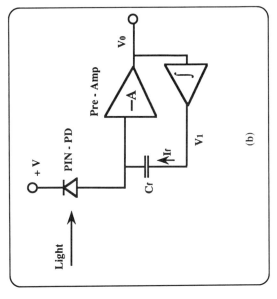

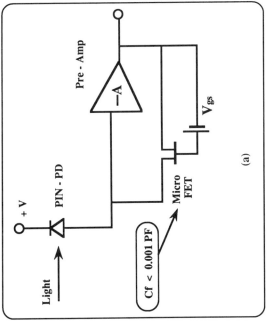

Figure 4.9 Active feedback design: (a) approach I; (b) approach II.

as a resistance, and the value is determined by the gate bias. Since the FET is fabricated in a small size, its capacitance C_f is less than 0.001 pF. Therefore, the merits of TZ design remain for a large R_f value. In approach 2, the feedback current I_f is made from an output voltage V_0 through an integrator, as shown in Figure 4.9(b). That is,

$$I_f = C_f \frac{d}{dt} V_1 = C_f \frac{d}{dt} \int V_0 \, dt = C_f V_0 \qquad (4.9)$$

where V_1 is the voltage shown in Figure 4.9(b). Equation (4.9) shows that the combination of a capacitance C_f and an integrator acts as an ideal resister. Experimentally, a 45-Mb/s receiver sensitivity of -51.7 dBm for a 10^{-9} error rate at a 1.3-μm wavelength and its dynamic range of 54 dB in optical level with the aid of the dynamic-range-extender circuit (AGC) were obtained.

Optical Feedback Design

In optical feedback design, a light path is used as a feedback loop path, instead of an electrical loop path, to eliminate feedback resistance. Therefore, thermal noise due to the feedback resistance R_f is removed, resulting in high sensitivity. The capacitance C_f is also removed and higher bandwidth is expected. With the optical feedback method, a large feedback current is possibly obtained from low output voltage, and wider dynamic range is expected. The basic configuration of optical feedback design is shown in Figure 4.10. In a front-end, an LED or a laser diode is used as a light source to generate feedback light, and a pin-PD is used to receive this feedback light. The feedback current i_f generated from the received feedback light at a pin-PD is

$$i_f = K i_{\text{LED}} \qquad (4.10)$$

where i_{LED} is the current of an LED (or i_{LD} in the case of a laser diode), and K is the feedback coupling factor. The parameter K includes the efficiencies of an LED and pin-PD used for an optical feedback, and the coupling efficiency between an LED and a photodiode. The transfer function $Z(f)$ is

$$Z(f) = \frac{R_s}{K} \frac{1}{1 + j2\pi f R_s C_T / AK} \qquad (4.11)$$

where R_s is the resistor shown in Figure 4.10 and C_T is the total capacitor, which includes both photodiode capacitances, an input capacitance of a preamplifier, and

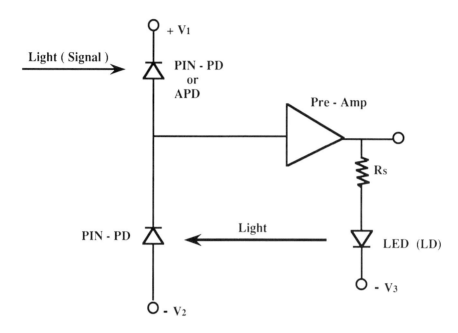

Figure 4.10 Optical feedback design.

a stray input capacitance. By comparing (4.6) and (4.11), the equivalent feedback resistance (R_f) is obtained at low frequency:

$$(R_f)_{eq} = R_s/K \tag{4.12}$$

The thermal noise from a resistor R_s is

$$\langle i_t^2 \rangle_{LED} = \frac{4kT}{R_s} B \tag{4.13}$$

and this noise is coupled to the input by optical feedback. By using (4.10), the thermal noise is

$$\langle i_t^2 \rangle = \frac{4kTK^2}{R_s} B \tag{4.14}$$

and this noise is compared to the noise in ordinary TZ design. By using the equivalent feedback resistance $(R_f)_{eq}$, the noise in the ordinary TZ design is

$$\langle i_t^2 \rangle_{eq} = \frac{4kT}{(R_f)_{eq}} B = \frac{4kTK}{R_s} B \tag{4.15}$$

The noise using optical feedback design is smaller than that using ordinary TZ design by a factor of K. When K is sufficiently small, the thermal noise in optical feedback design is much smaller and high sensitivity is expected.

The measured sensitivity of a receiver using optical feedback is -62.8 dBm at a 1.55-μm wavelength for 1.5 Mb/s, and its dynamic range is 40 dB [7]. In this experiment, InGaAs pin-PDs and a silicon *junction FET* (JFET) preamplifier were used. The feedback coupling ratio K was about 10^{-5}. The sensitivity of -62.8 dBm at a 1.55-μm wavelength for 1.5 Mb/s corresponds to 2650 photons/bit. When compared to the ordinary receiver using an InGaAs pin-PD, shown in Table 3.2, this value is very small. In [7–9], experiments using optical feedback have been done up to 34 Mb/s and have been restricted to relatively low bit rate transmission applications. The reason for this restriction originates in the closed-loop time constant in an optical feedback loop.

4.2.3 Low-Impedance Design

The approach of LZ design is different from TZ or HZ designs. In LZ design, load resistance R_L is designed to have low impedance. This design seeks simplicity and economy. The configuration of a receiver in LZ design is shown in Figure 4.11. Equations (4.2) and (4.3) also hold in this design. Without the EQL circuit, the necessary bandwidth must be realized in this design. Therefore, the value of R_L must be satisfied as follows:

$$R_L \leq \frac{1}{2\pi f_c C_T} \tag{4.16}$$

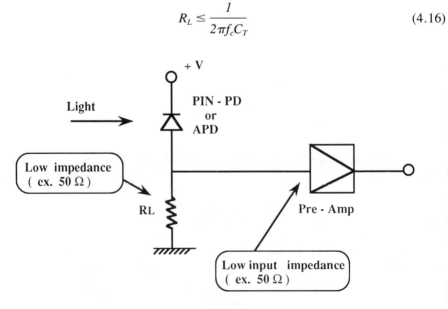

Figure 4.11 LZ design.

and the cutoff frequency f_c is around the value of B for digital transmission (B is the bit rate).

Thermal noise is expressed by (4.1) and takes a large value because of the low R_L value. Although an LZ-designed receiver has the lowest sensitivity among HZ, TZ, and LZ designs, wide bandwidth and the widest dynamic range are obtained easily. A commercially available amplifier can be used when designing $R_L = 50\ \Omega$ or $R_L = 75\ \Omega$. This may be the economical construction of a receiver and a system.

4.2.4 Resonance-Type Design for Coherent Transmission

Resonance-type design is intended for coherent heterodyne transmission systems [10]. In heterodyne transmission, high S/N is required only around the IF band. Therefore, a tuning circuit is inserted as shown in Figure 4.12. In this figure, one example of a tuning circuit is shown. For the case of no tuning circuit, a schematic S/N curve as a function of frequency is shown in the same figure as curve A. With the tuning circuit, a peak of the S/N curve moves to the IF band as indicated with curve B.

This approach is similar to the conventional radio heterodyne receiver and is applicable to a receiver design in optical SCM transmission systems.

4.2.5 Induction Peaking Design for High-Speed Transmission

Peaking design is known for conventional electric amplifier designs, and this technique is also applicable to a receiver required for wide bandwidth [11–15]. As we already discussed, the effect of C_T decreases the bandwidth of the front-end, where C_T is the total capacitor, which includes a capacitor C_d of a pin-PD or an APD, an input capacitor C_a, and a stray capacitor C_s. To compensate for this effect, the peaking technique has been investigated for analog transmission [12] and for high-speed transmission [13–15]. An example of inductor peaking and its effect is shown in Figure 4.13. Although simply an inductance is used in this figure, an actual circuit is complex when used for high-speed transmission [15]. Without peaking, transimpedance has a somewhat narrow bandwidth, and it is shown in Figure 4.13 as the curve A. The compensation as a result of peaking widens the bandwidth, and curve A moves to curve B. With inductor peaking, bandwidth improvement is estimated by a factor of 20% to 50% [15].

4.3 COMPARISON OF FRONT-END DESIGNS

The configurations, advantages, and drawbacks of the many receiver designs are summarized in Table 4.1. Except for dynamic range, these have already been explained.

As for dynamic range, wider dynamic range is obtained when a receiver can detect both low-power and high-power input signals without errors. This is difficult

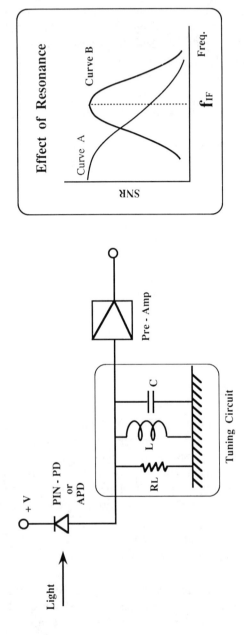

Figure 4.12 Resonant receiver design.

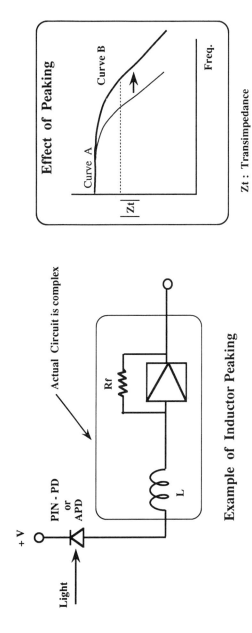

Figure 4.13 Peaking design.

Table 4.1
Comparison of Optical Receiver Designs

Receiver	Configuration	Advantage	Drawback
LZ	Low impedance (50 Ω termination)	Widest dynamic range Wide bandwidth	Low sensitivity
HZ	High impedance termination	Highest sensitivity	Narrow dynamic range Precise design for EQL
TZ	Transimpedance (feedback)	Wide dynamic range	
Resonance	Resonant receiver for heterodyne	SNR improvement at IF frequency	
Peaking	Inductor or capacitor peaking	Wide bandwidth by peaking (for high bitrate)	

for a receiver with high sensitivity, and there is a tradeoff between sensitivity and dynamic range. This is explained by Figure 4.14, where receiver configurations of three designs, LZ, HZ, and TZ, are shown. The upper receiving range for high light power signals is determined by the saturation voltage V_{sat}, as indicated in the figure. The current i_P is

$$i_P = RP \tag{4.17}$$

where R is the responsivity and is defined in (2.36), and P is the input light power. V_{sat} is obtained for each case as

$$V_{sat} = A i_P R_L = ARPR_L \quad \text{(LZ and HZ)} \tag{4.18}$$

$$V_{sat} = i_P R_f = RPR_f \quad \text{(TZ)} \tag{4.19}$$

Therefore, the upper receiving power P_{sat} is

$$P_{sat} = \frac{V_{sat}}{ARR_L} \quad \text{(LZ and HZ)} \tag{4.20}$$

$$P_{sat} = \frac{V_{sat}}{RR_f} \quad \text{(TZ)} \tag{4.21}$$

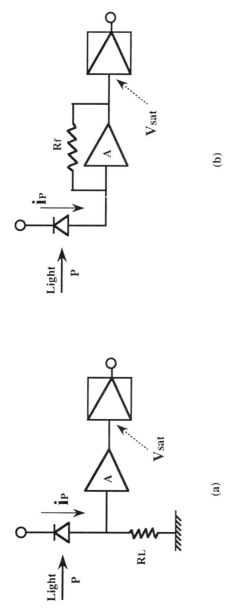

Figure 4.14 Sensitivity and dynamic range: (a) HZ or LZ design; (b) TZ design.

R_L tends to be a large value in HZ design, and so P_{sat} becomes a small value. This indicates that higher sensitivity means narrow dynamic range. This relation also holds in TZ design, as indicated by (4.8). Higher R_f value means higher sensitivity and narrow dynamic range. In LZ design, R_L takes a small value (e.g., 50 Ω), resulting in the widest dynamic range and the lowest sensitivity among these designs.

A system designer must select a suitable receiver design among the listed designs in Table 4.1. In ordinary digital transmission systems, TZ design is often selected. Its sensitivity and wide dynamic range are not so bad. For some systems, LZ design may be selected for economical reasons or simplicity, or if there is no need for high sensitivity.

4.4 AGC AND TIMING EXTRACTION

Although the front-end design is the most important for an optical receiver design, AGC is also important for the receiver design. AGC technology used in a conventional electric receiver is also applicable; however, the unique AGC for an optical receiver using APD has been developed. The same technologies of timing extraction developed for a conventional electric receiver is also applicable to an optical receiver in digital transmission systems.

4.4.1 AGC

AGC for a pin-PD receiver is shown in Figure 4.15, and this configuration is the same as that of a conventional electric receiver. The peak output voltage is detected by a peak detector (peak det.), and the detected voltage controls the gain of a voltage-controllable amplifier. When a larger peak value is detected, the gain is made smaller. One example of an AGC circuit is shown in this figure. The peak voltage is detected and held by a simple configuration of an electric diode D (ordinary diode, not a photodiode) and a capacitor C. The detected voltage controls an FET, which serves as voltage control resistance. The combination of this FET and an amplifier consists of a gain-controllable amplifier by the peak voltage.

AGC for an APD receiver is shown in Figure 4.16, and this configuration is unique for an optical receiver. In this configuration, two types of AGC are used, such as an ordinary electric AGC and a bias-voltage-controlled APD. The former AGC is the same as the AGC used in a pin-PD receiver. The latter is unique and uses the effect in which multiplication M is varied by the bias voltage. The variable M value is a gain controllable-amplifier by the peak voltage. In Figure 4.16, two types of AGC are used for an APD receiver. For some applications, one type of AGC (either of them) may be adequate.

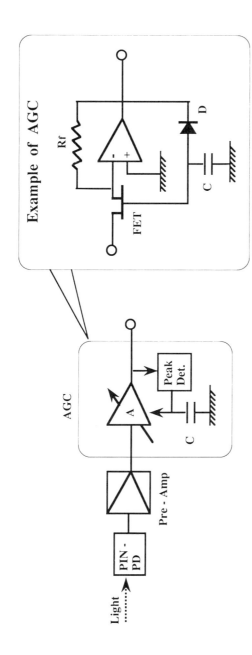

Figure 4.15 AGC for pin-PD receiver.

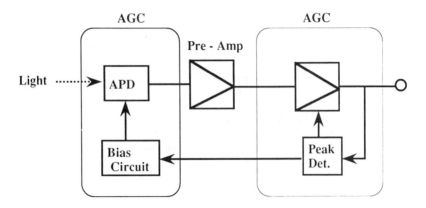

Figure 4.16 AGC for APD receiver.

4.4.2 Timing Extraction

Timing extraction in optical receivers in digital systems is the same as the conventional technology. Many timing extraction techniques have been developed, such as a resonance method using several tank circuits, a PLL method, and a digital sampling and a clock selection method. These are selected by considering the characteristics of transmitted signals. The characteristics of transmitted signals are the following:

1. Method of transmission: continuous or in bursts;
2. Number of strings of no-transition signal symbols;
3. Stability; and
4. Cost.

The selected method must be suitable for circuit integration (IC or LSI) for mass production and cost reduction.

4.5 DEGRADATION FACTORS

Basic noises, such as shot noises due to input light, dark current, thermal noise due to circuits, were discussed in the previous chapter. Some degradation factors, which are peculiar to the system, were also treated in the previous chapter. Here, degradation factors in receiver sensitivity are summarized and listed in Figure 4.17 for convenience. Common factors, which must be considered in whatever optical systems are designed, and peculiar factors of the special systems are listed separately. Factors other than those concerning the optical amplifier have already been discussed. The optical amplifier is discussed in Chapter 5. Among the factors listed in

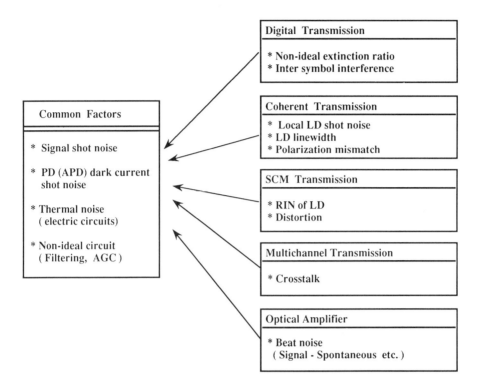

Figure 4.17 Degradation factors.

Figure 4.17, thermal noise in the common factors has a close relationship to front-end design. Some of the factors cannot be avoided by only receiver design or fabrication. They depend on the system parameters, such as optical frequency allocation and line code. Degradation also depends mainly on the device parameters, such as laser linewidth, RIN, and filtering characteristics. When some degradation in optical signal receiving is found, it is very important to know the origins of the degradation.

REFERENCES

[1] Muoi, T. V., "Receiver Design for High-Speed Optical-Fiber Systems," *IEEE J. of Lightwave Technol.*, Vol. LT-2, 1984, p. 243.
[2] Personick, S. D., "Receiver Design for Digital Fiber Optic Communication Systems, I," *B.S.T.J.*, Vol. 52, 1973, p. 843.
[3] Goell, J. E., "An Optical Repeater With High-Impedance Input Amplifier," *B.S.T.J.*, Vol. 53, 1974, p. 629.
[4] Ueno, T., Y. Ohgushi, and A. Abe, "A 40 Mb/s and a 400 Mb/s Repeater for Fiber Optic Communications," *Proc. European Conf. Optical Commun. (ECOC'75)*, 1975, p. 147.

[5] Hullett, J. L., and T. V. Muoi, "A Feedback Receiver Amplifier for Optical Transmission Systems," *IEEE Trans. Commun.*, Vol. COM-24, 1976, p. 1180.
[6] Williams, G. F., and H. P. Leblanc, "Active Feedback Lightwave Receivers," *IEEE J. of Lightwave Technol.*, Vol. LT-4, p. 1502. 1986,
[7] Kasper, B. L., A. R. McCormik, C. A. Burrus, and J. R. Talman, "An Optical-Feedback Transimpedance Receiver for High Sensitivity and Wide Dynamic Range at Low Bit Rates," *IEEE J. of Lightwave Technol.*, Vol. 6, 1988, p. 329.
[8] Methley, S. G., "-70 dBm APD Optical Feedback Receiver at 2.048 Mb/s," *Electron. Lett.*, Vol. 24, 1988, p. 1259.
[9] Hunter, C. A., R. L. Calton, P. R. Sadd, and S. D. Walker, "Optical Feedback Receiver With Sensitivity of -55.2 dBm at 8 Mb/s," *Electron. Lett.*, Vol. 27, 1991, p. 886.
[10] Kikuch, K., T. Okoshi, and K. Emura, "Achievement of Nearly Shot-Noise-Limited Operation in a Heterodyne-Type PCM-ASK Optical Communication System," *Proc. European Conf. Optical Commun. (ECOC'82)*, Cannes, France, No. AXII-6, 1982, p. 419.
[11] Hulett, J. H., and T. V. Muoi, "Modified Receiver for Digital Optical Fiber Transmission Systems," *IEEE Trans. Commun.*, Vol. COM-23, 1975, p. 1518.
[12] Nagano, K., Y. Takahashi, Y. Takasaki, M. Maeda, and M. Tanaka, "Optimizing Optical Transmitting Multichannel Video Signals Using Laser Diodes," *IEEE Trans. Commun.*, Vol. COM-29, 1981, p. 41.
[13] Ohkawa, N., J.-I. Yamada, and K. Hagimoto, "Broadband GaAs FET Optical Front-End Circuit up to 5.6 GHz," *Electron. Lett.*, Vol. 22, 1986, p. 259.
[14] Gimlett, J. L., "Low-Noise 8 GHz PIN/FET Optical Receiver," *Electron. Lett.*, Vol. 23, 1987, p. 281.
[15] Ohkawa, N., "Fiber-Optic Multigigabit GaAs MIC Front-End Circuit With Inductor Peaking," *IEEE J. of Lightwave Technol.*, Vol. 6, 1988, p. 1665.

Chapter 5
Optical Amplifier

In this chapter, optical amplifiers—including a semiconductor laser diode amplifier and a fiber amplifier—are discussed. Optical amplification is applicable to all the transmission schemes examined in Chapter 3. Among them, optical amplification is very useful in IM/DD in lowering the contribution of thermal noise of circuits and achieving high receiver sensitivity. It is also useful in the video distribution system in compensating for distribution loss. For WDM and OFDM systems, the characteristics of simultaneous amplification are very important and are also discussed.

5.1 OUTLINE OF OPTICAL AMPLIFIER

There are many types of lasers, such as gas, solid-state, semiconductor, and fiber Raman lasers. Without amplification, lasing cannot occur. Therefore, a laser can be modified to be an optical amplifier in principle. Many types of optical amplifiers have been proposed and investigated. The classification of optical amplifiers for optical transmission is shown in Figure 5.1. In this figure, a *semiconductor laser diode amplifier*, abbreviated LD amp or SLDA, uses a semiconductor laser [1–3]. Two types of LD amp, the *Fabry-Perot Amplifier* (FPA) and the *traveling wave amplifier* (TWA), are considered. The doped-fiber amplifier uses an optical fiber with a special dopant [4–7]. The EDFA can amplify light at a wavelength of around 1.5 μm, and this is one of the wavelengths used in the optical transmission systems [8–11]. A laser diode has been proposed as a pumping source [12], and this shows the EDFA to be an attractive optical device for optical transmission systems. Other dopants (N_d^{3+}, Sm^{3+}, Tu^{3+}, Pr^{3+}) in doped-fiber amplifiers have been investigated to meet the demand for other wavelength amplifications, such as a 1.3-μm wavelength [11,13,14] and around a 1.7-μm wavelength [15,16]. Using the fiber nonlinearity as discussed in Chapter 3, optical amplification is possible, and a fiber Raman amplifier and a Brillouin amplifier are realized [17–19].

Among these optical amplifiers, the LD amp, the doped-fiber amplifier, and the fiber Raman amplifier are used for optical transmission experiments. In Figure

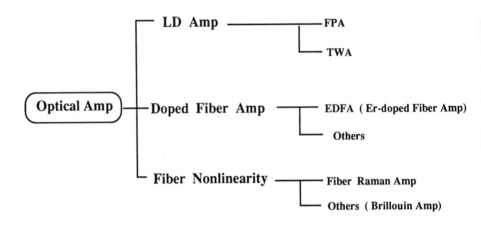

Figure 5.1 Classification of optical amplifiers.

5.2, the configurations of these amplifiers are shown. An LD amp is pumped by a current, while a doped-fiber amplifier and a fiber Raman amplifier are pumped by a light (a laser light). Two types of LD amps, FPA and TWA, are considered and are shown in Figure 5.3. They are classified by operating principle, a single path or multiple paths in the gain medium (a laser chip). In the FPA, multiple reflections take place at LD facets and the resonance is formed. Therefore, gain profile is just like a Fabry-Perot resonator, as shown in this figure. There are many peaks and bottoms. In general, this type of LD amp uses a laser diode without any modification. In the TWA, no multiple reflections take place and a single path is realized with very small reflectivity at LD facets (ideally no reflection). To realize low reflectivity, a special optical coating on LD facets is applied. The gain profile of a TWA is also shown in this figure, and there is one broad peak. The advantage of a laser diode is the small dimensions, and it can be integrated into a photonic IC with other devices. The EDFA, one example of doped-fiber amplifiers, is attractive because of the possibility of LD pumping and its superior characteristics, which are explained later. The fiber Raman amplifier cannot be pumped by a laser diode at present. This is a disadvantage for optical transmission applications. The same pumping direction and the opposite pumping direction for the signal light direction are possible for doped-fiber and fiber Raman amplifiers.

Several applications for optical amplifiers can be considered and applications for an optical transmission are shown in Figure 5.4. In Figure 5.4(a), optical amplifiers are used as a postamplifier, an inline amplifier, and a preamplifier. A post

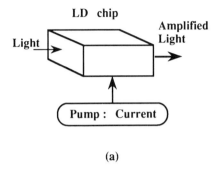

(a)

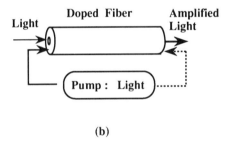

(b)

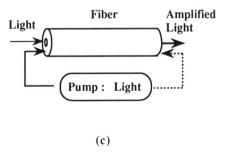

(c)

Figure 5.2 Configurations of typical optical amplifiers: (a) LD amp; (b) doped-fiber amp; (c) fiber Raman amp.

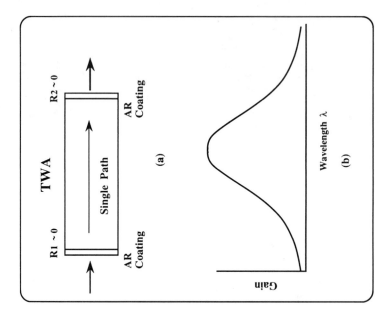
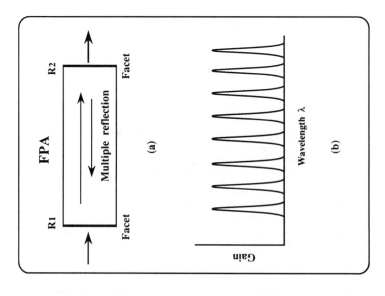

Figure 5.3 Fabry-Perot and traveling wave amplifiers: (a) structure; (b) character.

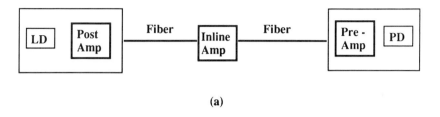

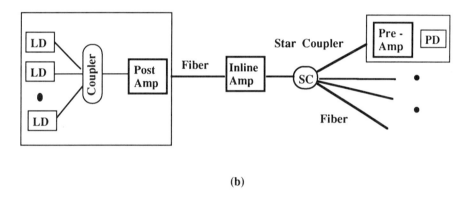

Figure 5.4 Applications of optical amplifiers: (a) as a postamplifier, an inline amplifier, and a preamplifier; (b) in a PON.

amplifier is used to increase the sending light power in the case of a lack of light source power. An inline amplifier is used to compensate the fiber cable loss and serves as an analog repeater. A preamplifier is used to increase the input light power and its main purpose is to decrease the contribution of thermal noise of circuits. Optical amplifiers are also applied in a PON, and they are shown in Figure 5.4(b). In the case of multiple laser diodes used in the system, lights from these lasers are combined by a coupler and they may be amplified by a postamplifier. An inline amplifier is used to compensate for the fiber cable loss and the branching loss of a star coupler. A 16-branch star coupler has about 12 dB of loss in the case of no insertion loss. The configuration shown in Figure 5.4(b) is useful for the video distribution system to compensate for the distribution loss.

Characteristics of typical optical amplifiers are shown in Table 5.1, and typical values are listed in this table for comparison. The gain, bandwidth, and polarization dependence are important for any optical applications. For all optical amplifiers, about 30 dB of gain is possible, and the gain depends on the pumping current or the pumping light power. In a TWA, the necessary current for pumping is large when compared to the FPA because of its single-path nature. In this table, R in the TWA column indicates the residual reflectivity at LD facets because of incomplete coating. EDFA is more attractive than a fiber Raman amplifier because of less pumping light power, and it indicates the possibility of the use of a laser diode. A 46.5-dB EDFA gain was obtained under the conditions of Er dopant value = 100 ppm, Er-doped-fiber length = 100m, pumping wavelength = 1.48 μm, and pumping power = 130 mW [20]. Bandwidth is especially important for the application of WDM or OFDM systems. The FPA has a narrow bandwidth, and other systems have relatively large bandwidths. Saturation output power is an important factor for postamplification and sometimes inline amplification applications. A +11.3-dBm saturation power of EDFA with a 37.1-dB gain was realized with the conditions of Er-doped-fiber length = 47.5m and pumping power = 53.6 mW [21]. Generally, optical amplifiers using fibers have no polarization dependence, while LD amps do. Recent investigation of the TWA succeeded in realizing no polarization dependence or slight polarization dependence in several types of TWAs [22–25]. The characteristics listed in this table are explained in detail in the following sections.

Table 5.1
Comparison of Typical Optical Amplifiers

	LD Amp		Doped-Fiber Amp (EDFA)	Fiber Nonlinearity (Fiber Raman)
	TWA	FPA		
Typical gain	~30 dB (R = 0.01%)	20 ~ 30 dB	30 ~ 45 dB	20 ~ 45 dB
Bandwidth (20-dB gain)	~1,000 GHz	1 ~ 10 GHz	~500 GHz	~3,000 GHz
Saturation output power	~+10 dBm	−10 ~ −5 dBm	+5 ~ +10 dBm	~ +30 dBm
Polarization dependence	Yes (No)	Yes	No	No
Pumping (Typ)	Current (100 ~ 200 mA)	Current (~20 mA)	LD light (20 mW ~ 100 mW)	Laser light (1 ~ 5 W)

5.2 SEMICONDUCTOR LASER DIODE AMPLIFIER

The gain characteristics of the laser diode amplifier, such as gain profile, saturation power, and polarization dependence, are explained here.

5.2.1 Gain Profile of Laser Diode Amplifier

The gain profile of an LD amp depends on the reflection characteristic (a single path or multiple paths) of a laser cavity. This is analyzed using a Fabry-Perot resonance model, as shown in Figure 5.5. Two LD facets make up the Fabry-Perot resonator. Incident light enters an FP resonator from the left side (LD facet) in Figure 5.5(a), and this light passes through the active layer with a gain and then travels out from the right side of the resonator (LD facet). A portion of the light reflects at the right side and then passes through the active layer and again reflects. This reflected light passes through the active layer and travels out from the right side. Some portion of this double-path light reflects at the right side and passes through the active layer, and so on. These multiple reflections are expressed as follows:

The first transmitted light: $E_i t_1 t_2 e^{-\gamma L}$,
The second transmitted light: $E_i t_1 r_1 r_2 t_2 e^{-3\gamma L}$; and so on.

The notations E_i, t_1, t_2, r_1, r_2, γ, and L are the electric field of incident light, the transmittance coefficient at the left LD facet, the transmittance coefficient at the right LD facet, the reflection coefficient at the left LD facet, the reflection coefficient at the right LD facet, the complex propagation constant, and the laser diode length, respectively. The field of amplified output light from an LD amp is the sum of these transmitted lights and it is

$$E_t = E_i t_1 t_2 e^{-\gamma L}[1 + r_1 r_2 e^{-2\gamma L} + (r_1 r_2 e^{-2\gamma L})^2 + \ldots]$$
$$= \frac{t_1 t_2 e^{-\gamma L}}{1 - r_1 r_2 e^{-2\gamma L}} E_i \quad (5.1)$$

and the power transmittance G_T is

$$G_T = \frac{|E_t|^2}{|E_i|^2} = \frac{|t_1 t_2 e^{-\gamma L}|^2}{|1 - r_1 r_2 e^{-2\gamma L}|^2}. \quad (5.2)$$

The propagation constant γ is

$$\gamma = -(\Gamma g - \alpha)/2 + j\beta \quad (5.3)$$

where, Γ, g, α, and β are the mode confinement factor, the optical gain in active layer, the optical loss in active layer, and the phase constant, respectively.

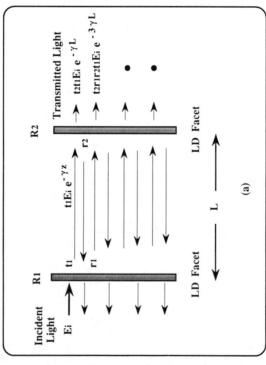
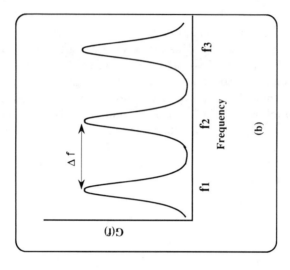

Figure 5.5 Fabry-Perot resonance model: (a) FP resonator; (b) gain profile.

The phase constant β is $2\pi n/\lambda$, where n is the refractive index of the active layer. Using (5.3), the calculated result of (5.2) is

$$G_T = \frac{(1 - R_1)(1 - R_2)G_s}{(1 - \sqrt{R_1 R_2}\, G_s)^2 + 4\sqrt{R_1 R_2}\, G_s \sin^2(\beta L)} \qquad (5.4)$$

where the following relations between reflection coefficients for amplitude and reflection coefficients for power are used:

$$R_1 = r_1^2 \qquad R_2 = r_2^2 \qquad (5.5)$$

And G_s is defined as

$$G_s = \exp[(\Gamma g - \alpha)L] \qquad (5.6)$$

and this is the unsaturated gain for a single path. Equation (5.4) is rewritten by using optical frequency f and the phase reference (phase = 0) taken at $f = f_1$ [26]:

$$G(f) = \frac{(1 - R_1)(1 - R_2)G_s}{(1 - \sqrt{R_1 R_2}\, G_s)^2 + 4\sqrt{R_1 R_2}\, G_s \sin^2[2\pi(f - f_1)L/V_g]} \qquad (5.7)$$

where V_g is the group velocity inside the cavity. The gain peak exists at $f = f_1$ and other gain peaks exist because of the periodic character of (5.7) and are shown in Figure 5.5(b). The frequency interval Δf between peaks is obtained by (5.7) as

$$\Delta f = V_g/(2L) \qquad (5.8)$$

and the bottom near f_1 is located at $f = f_1 + \Delta f/2$. The maximum gain (peak gain) and the minimum gain (bottom gain) are obtained by using (5.7), and they are

$$G_{\max} = \frac{(1 - R_1)(1 - R_2)G_s}{(1 - \sqrt{R_1 R_2}\, G_s)^2} \qquad (5.9)$$

$$G_{\min} = \frac{(1 - R_1)(1 - R_2)G_s}{(1 + \sqrt{R_1 R_2}\, G_s)^2} \qquad (5.10)$$

The ratio of peak gain and bottom gain approaches 1 when

$$\sqrt{R_1 R_2}\, G_s \ll 1 \qquad (5.11)$$

A laser amplifier, where (5.11) is satisfied, is a TWA. To satisfy this condition, R_1, R_2, or G_s must be small. Usually, G_s is set to be large, then R_1 and R_2 must take very small values. In an ideal case, the values are zero. Gain ripples, as shown in Figure 5.5(b), do not occur even for the infinite G_s value in this ideal case. In a TWA, gain bandwidth is broad because there are no large gain ripples. This characteristic is very attractive for WDM or OFDM applications.

5.2.2 Saturation Output Power

The carrier and photon inside a laser diode are described by the rate equations (2.27) and (2.28). In the case of an LD amp, an optical injection term is added to these equations [27]:

$$\frac{dN}{dt} = \frac{I}{eV} - \frac{N}{\tau_c} - A_g V_g (N - N_0) S \tag{5.12}$$

$$\frac{dS}{dt} = A_g V_g \Gamma (N - N_0) S - \frac{S}{\tau_p} + \beta \Gamma \frac{N}{\tau_c} + S_{in} \tag{5.13}$$

where N, S, I, e, V, A_g, N_0, V_g, and β are the carrier density inside the cavity, the photon density inside the cavity, the injected current, the electron charge, the volume of active layer, the differential gain coefficient, the carrier density in transparent condition (the same as N_0 in (2.22)), the group velocity inside the cavity, and the spontaneous emission factor (which is the fraction of spontaneous emission entering the lasing mode), respectively. Γ is the mode confinement factor (which is the factor of optical mode power inside an active layer). τ_c and τ_p are the carrier lifetime and the photon lifetime, respectively. S_{in} is the injected photon density, and this is the added optical injection term which corresponds to the input light to an LD amp. To determine the gain, the carrier density in the steady state is obtained by (5.12):

$$\frac{I}{eV} = \frac{N}{\tau_c} + A_g V_g (N - N_0) S \tag{5.14}$$

N is then obtained as

$$N = \frac{(I/eV) + A_g V_g N_0 S}{(1/\tau_c) + A_g V_g S} \tag{5.15}$$

and the gain coefficient per unit length g is

$$g = A_g (N - N_0) \tag{5.16}$$

Using (5.15) and (5.16), g is

$$g = A_g \frac{(I/eV)\tau_c - N_0}{1 + A_g V_g \tau_c S} = \frac{g_0}{1 + BV_o S} \quad (5.17)$$

where g_0 and β are

$$g_0 = A_g[(I\tau_c/eV) - N_0] \quad (5.18)$$

$$B = A_g V_g \tau_c / V_o \quad (5.19)$$

and V_o is the optical volume and is V/Γ. S in (5.17) is the steady-state solution of (5.12) and (5.13). Since the intensity of light is $I_{op}(z) = hfV_g S(z)$, then (5.17) is rewritten as [28]

$$g(z) = \frac{g_0}{1 + I_{op}(z)/I_s} \quad (5.20)$$

where $I_s = hf/(A_g\tau_c) = hfV_g/(V_o B)$. When $I_{op}(z) = I_s$, then gain g decreases to half of g_0. In the saturation situation, G_s is expressed as (5.21) instead of (5.6) [28]:

$$G_s = \exp\left[\int_0^L \{\Gamma g(z) - \alpha\} dz\right] \quad (5.21)$$

The saturation output power of the TWA LD amp is higher than that of the FP-LD amp, as shown in Table 5.1. Since a TWA LD amp operates at a high injection current level, the carrier density of the TWA is higher than that of FPA (FP-LD amp). Therefore, the carrier lifetime τ_c of a TWA is short when compared to τ_c of an FPA, and then I_s of the TWA becomes larger than that of the FPA. This explains the saturation output power difference.

An example of the measured wavelength characteristics of FPA gain is explained below [29]. An experimental setup and procedure are shown in Figure 5.6. First, a 1.3-μm InGaAsP FP-LD (1) was driven above its threshold and its spectrum was measured by an optical spectrum analyzer (Figure 5.6(a)). The measured spectrum (FP-LD (1)) is shown in Figure 5.7(a) as the shaded spectra. The laser is multimode lasing, and the number of lasing modes is six in this scale. They are labeled A through F. Although no mode shorter than A can be seen in Figure 5.7(a), mode G does exist and is shown as such in the figure. Next, the spectrum of the FP-LD (II) with bias current $I = 0.97I_{th}$ was measured (Figure 5.6(b)). The FP-LD (II) had one open face (rear face). The spontaneous emission spectrum is shown in Figure 5.7(a) as the unshaded spectra. The spectra form a continuous curve in this scale.

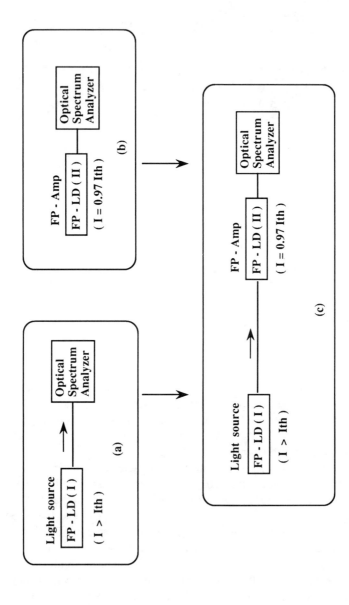

Figure 5.6 Experimental procedure for measuring the wavelength dependency of FP amp: (a) spectrum measured by optical analyzer; (b) spectrum of FP-LD II; (c) light output from the FP-LD II rear face.

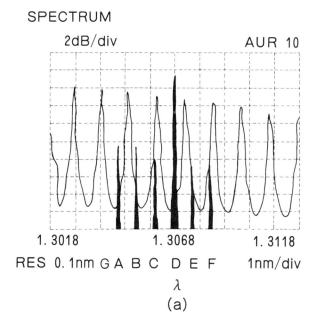

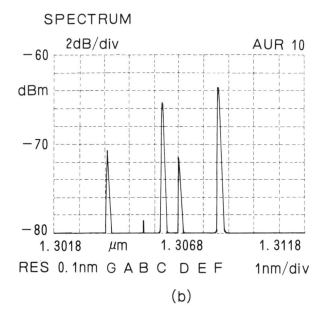

Figure 5.7 Measured optical spectrum of (a) Figure 5.6(a); (b) Figure 5.6(b) (© 1992 IEEE).

In this experiment, the FP-LD (II) was used as an optical amplifier. Finally, the output of light from the FP-LD (II) rear face in the Figure 5.6(c) configuration was measured. The bias current of FP-LD (II) was $0.97I_{th}$. The measured spectrum is shown in Figure 5.7(b). The output light level corresponds to the relations of the input laser spectrum (FP-LD (I)) and the spontaneous emission spectrum of the amplifier (FP-LD (II)). Low-level modes, such as mode G, are amplified because their wavelength coincides with the spontaneous emission peaks. Modes A and B result in low-level outputs because their wavelengths are located at spontaneous emission troughs. These FPA characteristics are not attractive for many applications. In the case of the TWA, the gain bandwidth is much broader, as indicated in Table 5.1.

5.2.3 Polarization Dependence of Gain

An example of the measured polarization dependence is shown below [29]. A DFB laser was used as the light source and it was used in the single-mode oscillation state at near 1317 nm. Its temperature was controlled. A third FP-LD module (FP-LD (III)) was used as a preamplifier. To change the polarization state of input light, a *polarization controller* (PC) was used. Polarization states were changed to determine maximum and minimum output from the PD, which is located at the rear of the preamplifier. The measured gains (MAX and MIN) are plotted as a function of the normalized bias current I/I_{th} of the FP-LD (III) and are shown in Figure 5.8. The threshold current I_{th} is 28.0 mA. The gain in Figure 5.8 is defined as follows:

$$\text{Gain} = \{P_0(\text{input}, I) - P_0(\text{input} = 0, I)\}/P_0(\text{input}, I = 0) \qquad (5.22)$$

where P_0 (input, I), P_0 (input = 0, I), and P_0 (input, $I = 0$) are the output powers with input light and bias current I, with no input light and bias current I, and with input light and no bias current, respectively. P_0 (input = 0, I) represents the spontaneous emission and P_0 (input, $I = 0$) reflects the preamplifier (FP-LD (III)) loss at zero bias. The results indicate that the gain difference is very small when $I/I_{th} <$ 0.6, and the gain itself is small (about 2). In Figure 5.8, the spontaneous emission is also plotted in arbitrary units. By changing the DFB laser temperature, the gains (MAX, MIN) were measured as a function of wavelength λ. The results are shown in Figure 5.9. At low bias currents, the polarization difference of gain is very small and the gain itself is also small.

It is well known that ordinary *double heterostructure laser diodes* (DH LD) have the *transverse electric* (TE) lasing mode. The LD facet reflectivity difference for TE and *transverse magnetic* (TM) modes are explained in [30–32]. Both reflectivity R and mode confinement factor Γ are different for TE and TM modes. This explains that both the FPA and ordinary TWA of laser amplifiers have different gain values in TE and TM modes. Figures 5.8 and 5.9 are examples in the case of the

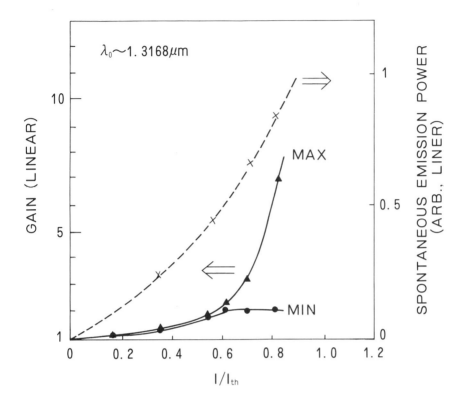

Figure 5.8 Maximum and minimum gain as a function of normalized bias current (© 1992 IEEE).

FPA. It was also measured that the wavelength for TE mode peak gain is different from that for TM mode peak gain [33]. This can also be seen in Figure 5.9(a,b).

Recently, TWAs with slight or no polarization dependence were reported at laboratory level [22–25]. To obtain these TWAs, four approaches have been taken:

1. Lowering the facet reflectivity R;
2. Equaling the mode confinement factors Γ for TE and TM modes;
3. Enhancement of TM mode gain;
4. Configuration.

The first three approaches are shown schematically in Figure 5.10. By lowering the facet reflectivity R, there is little difference of R for TE and TM modes. *Antireflective* (AR) coating at the laser facet was proposed for making TWAs, and TWAs with AR coating have been successfully fabricated. However, they were not sufficient for polarization-independent TWAs. The facet reflectivity R on the order of 10^{-4} was realized by using a window facet structure [22,24] and a buried facet structure [23]. The difference of the mode confinement factors Γ in TE and TM modes

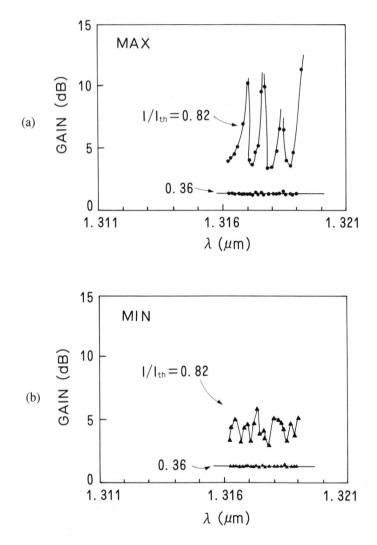

Figure 5.9 Measured gain as a function of wavelength: (a) maximum; (b) minimum (© 1992 IEEE).

originates in the asymmetric active layer structure. Typical value of cross-sectional dimensions are 0.15×2 μm. To decrease this asymmetric active layer structure, a thick active layer was realized [23,24]. By taking the combination of the first two approaches, TWAs with a TE-TM mode gain difference of about 1 dB or less have been developed at the laboratory level [22–24]. For the third approach, a strained *multiple quantum well* (MQW) has been proposed to enhance the TM mode gain

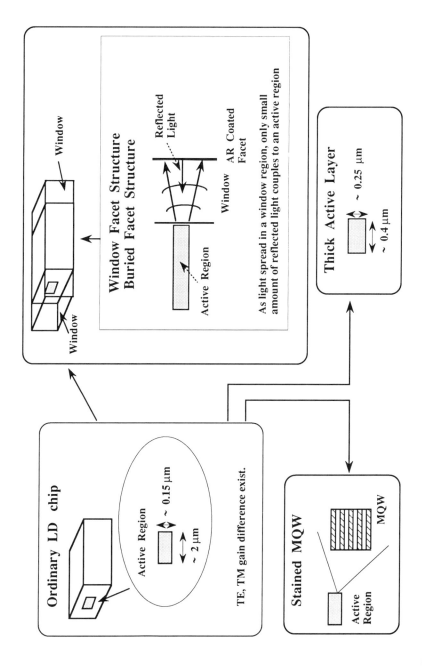

Figure 5.10 Polarization-independent TWA.

[25]. Although the obtained gain is 7.5 dB and small, a TWA with no gain difference was obtained [25]. The same approach was also applied for laser fabrication, and both a TM mode lasing FP laser and a TM mode lasing DFB laser have been developed [34]. The fourth approach uses ordinary TWAs. One example of this approach is to use two amplifiers in parallel, one for each polarization [35]. Another example is to use one amplifier and the double optical pass through the same amplifier, and a Faraday rotator is inserted in the pass [36]. The disadvantage of the fourth approach is its complexity.

A strained MQW is introduced to the TWA fabrication for polarization-independence. Recently, an MQW structure has been adopted for improving the TWA character, and some values in Table 5.1 may be altered. For example, the broad bandwidth of about 10,000 GHz was obtained [37].

5.3 DOPED-FIBER AMPLIFIER

The EDFA has been intensively investigated and its character is superior among doped-fiber amplifiers. Therefore, the EDFA is considered here as an example of a *doped-fiber amplifier* (DFA). Many characteristics of the EDFA are also applicable to other DFAs.

5.3.1 Configuration

The typical configuration of the EDFA is shown in Figure 5.11. At the initial stage of investigation of the EDFA, an argon-ion laser or a dye laser was used as a pumping source. However, a high-power laser diode is preferable as a pumping source for practical applications [38]. Three-level and four-level models for the DFA are shown in Figure 5.12. The EDFA is considered to be a three-level model. In the case of a three-level system, the final state is equal to the initial state (ground state). Therefore, the absorption of the ground state exists. This results in absorption for the weak pump power case. Amplification occurs when the pump power is above the certain value (the threshold pump power). Since the EDFA belongs to a three-level system, it has the optimum doped-fiber length for amplification. For longer length, the pump power becomes weak, and amplified signal light is absorbed at that part. A N_d^{3+}-doped-fiber amplifier is considered to be a four-level model. In the case of a four-level system, the final state is an intermediate state and is different from the initial state (ground state). Therefore, there is no absorption of ground state and no pump threshold.

Reported wavelengths for LD pumping are 0.82 μm [39], 0.98 μm [40], and 1.48 μm [38]. The reported gain coefficients for a pump power are 0.3 ~ 0.4 dB/mW for 0.82 μm, 4 ~ 5 dB/mW for 0.98 μm, and 2 ~ 4 dB/mW for 1.48 μm wavelength. Pumping of 0.98 μm has the highest gain coefficients among them.

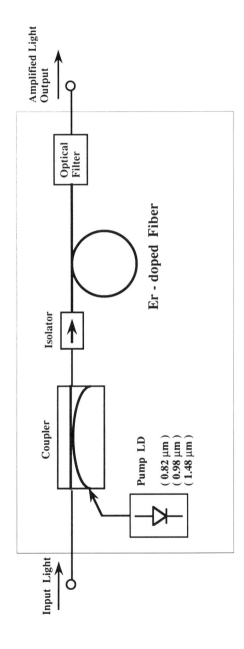

Figure 5.11 Configuration of EDFA.

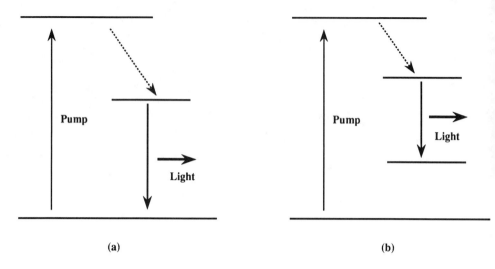

Figure 5.12 DFA models: (a) three-level model; (b) four-level model.

Possible wavelength ranges for pumping in these regions are 0.800 μm ~ 0.825 μm (25 nm), 0.975 μm ~ 0.985 μm (10 nm), and 1.450 μm ~ 1.485 μm (35 nm). Pumping of 1.48 μm has the widest wavelength range among them. Noise figures of an EDFA pumped by these wavelengths are also slightly different, and these are discussed later.

An isolator is used in Figure 5.11 in order to suppress the laser oscillation and also to prevent the feedback of *amplified spontaneous emission* (ASE). A fusion splice instead of an optical connector is preferable for connecting the components shown in this figure. High-performance optical connectors with sufficient high reflection loss must be applied when the connector is used. A countermeasure for the antireflection is very important for realizing a stable high gain amplification.

An optical filter near the output in Figure 5.11 is used to eliminate ASE. By using a narrowband optical filter, the EDFA shows good noise performance. Usually a bandwidth filter of a few nanometers is used. This optical filter inside the EDFA can be eliminated when a good noise performance is not required or when other filtering means outside the EDFA are possible.

5.3.2 Gain of EDFA

A typical gain profile of the EDFA is shown in Figure 5.13(a), and that of a TWA LD amp is shown in the same figure for comparison. The typical gain bandwidth of the EDFA is in the 20 ~ 30 nm range. When compared to a TWA LD amp, the

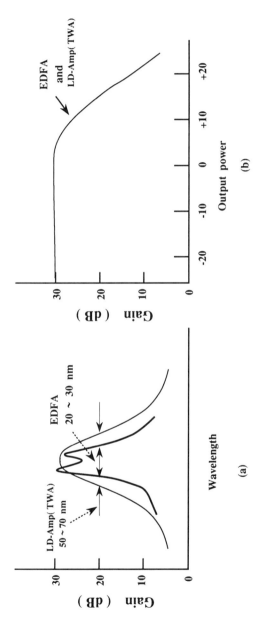

Figure 5.13 Gain characteristics of EDFA: (a) typical gain profile; (b) typical saturation output power.

EDFA's gain bandwidth is relatively small and the gain of the EDFA has two peaks near 1.535 and 1.55 μm. For OFDM or WDM applications, a flat gain against the wavelength is desirable. By codoping aluminum with erbium, an EDFA with a relatively flat and relatively wider bandwidth (35 nm) is obtained between the 1.525- and 1.56-μm wavelengths [41].

For obtaining the high gain coefficients for a pump power, the Er dopant concentration in the center part of the core is very important.

The reason? The intensity of pump light has the peak at the center of the core, according to the mode field profile of a single-mode fiber which has an approximate Gaussian profile. There is low pump light intensity near the vicinity of the core-cladding boundary when compared to the low pump light intensity near the core center. The signal may be absorbed near the boundary according to the three-level system if the pump light is weak. The matching of the Er dopant concentration location and the pump light intensity peak location causes the EDFA to have high gain coefficients.

5.3.3 Saturation Output Power

The gain of the EDFA is a single-path gain, which is the same situation of a TWA LD amp, and its gain G is expressed as

$$G = \exp\left[\int_0^L \{\Gamma g(z) - \alpha\} \, dz\right] \quad (5.23)$$

where

$$g(z) = \frac{g_0}{1 + I_{op}(z)/I_s} \quad (5.24)$$

g_0, $I_{op}(z)$, and I_s are the small signal gain, the intensity of light, and the saturation light intensity, respectively.

Although equation (5.24) is similar to equation (5.20), I_s in (5.24) is a function of pump power. Therefore, the saturation light intensity I_s increases with the pump light power. When $I_{op}(z) = I_s$, the gain becomes half of that at a small optical signal.

A typical saturation output power of both the EDFA and TWA LD amps is shown in Figure 5.13(b). When the values of I_s and g_0 for both the EDFA and TWA LD amps take similar values, the same curve in Figure 5.13(b) holds for both the EDFA and TWA LD amps.

5.3.4 Polarization Dependence of Gain and Coupling with a Fiber

Apart from an LD amp, the gain of the EDFA has no polarization dependence. This feature makes the EDFA very attractive for many applications. Another attractive feature of the EDFA is the low coupling loss between the EDFA and a single-mode fiber. Since the EDFA is composed of a doped single-mode fiber, it is relatively easy to realize a low-loss coupling with ordinary single-mode fibers. The coupling loss between a laser diode and single-mode fibers is larger than that in the EDFA case. When the coupling loss between a laser diode and a single-mode fiber is 3 dB, the fiber-to-fiber gain is less than its original gain by a factor of 6 dB in the ordinary configuration.

5.4 NOISE OF OPTICAL AMPLIFIER

5.4.1 Basic Noise Equation of Optical Amplifier

Here, the basic equation for calculating optical amplifier noise is derived by using the master equation [42–44]:

$$\frac{dP_n}{dt} = -[a(n+1) + bn]P_n + anP_{n-1} + b(n+1)P_{n+1} \tag{5.25}$$

where:

P_n = probability of state $|n>$ ($|n>$ is the state with n photons);
a = probability of spontaneous emission;
b = probability of stimulated absorption per photon.

This equation is derived by the following consideration. The probability of absorbing a photon in the unit time for one atom in the case of n photons is expressed as ζn, where ζ is the probability of the spontaneous emission. The probability of emitting a photon in the unit time for one atom is $\zeta(n + 1)$. These are the results from the quantum mechanical calculations. Here, we represent N_1 and N_2 as the atomic numbers for the ground state and the excited state, respectively. When n photons exist (its probability is P_n), the atoms at the ground state (N_1) absorb a photon with the probability of $\zeta n N_1$, and the atoms at the excited state (N_2) emit a photon with the probability of $\zeta(n + 1)N_2$. Therefore, P_n decreases by the amount of $[\zeta n N_1 + \zeta(n + 1)N_2]P_n$. P_n increases by amount $[\zeta n N_2 P_{n-1} + \zeta(n + 1)N_1 P_{n+1}]$ because the atoms at the excited state (N_2) for ($n - 1$) photons emit a photon (to transit to $|n>$) with the probability of $\zeta n N_2 P_{n-1}$, and the atoms at the ground state (N_1) for ($n + 1$) photons absorb a photon (to transit to $|n>$) with the probability of

$\zeta(n + 1)N_1 P_{n.+1}$. This situation is shown in Figure 5.14 by rewriting these probabilities with $\zeta N_2 = a$ and $\zeta N_1 = b$. The balance sheet for P_n is

$$-[a(n + 1) + bn]P_n + anP_{n-1} + b(n + 1)P_{n+1}$$

This results in (5.25).

By using the master equation, the average and square mean values for photons are obtained. These are expressed by

$$\langle n \rangle = \sum_n nP_n \qquad (5.26)$$

$$\langle n^2 \rangle = \sum_n n^2 P_n \qquad (5.27)$$

By multiplying n and n^2 by (5.25) and summing with respect to n, we obtain [43]

$$\frac{d\langle n \rangle}{dt} = (a - b)\langle n \rangle + a \qquad (5.28)$$

$$\frac{d\langle n^2 \rangle}{dt} = 2(a - b)\langle n^2 \rangle + (3a + b)\langle n \rangle + a \qquad (5.29)$$

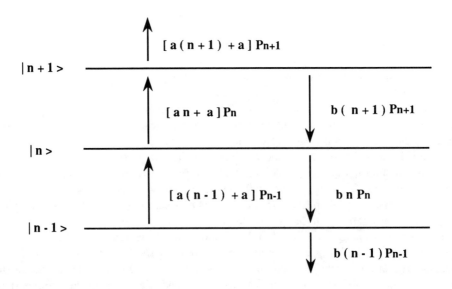

Figure 5.14 State transition (derivation of master equation).

By solving (5.28) and (5.29) with the initial conditions $\langle n \rangle = \langle n_0 \rangle$ and $\langle n^2 \rangle = \langle n_0^2 \rangle$ at $t = 0$, the following results are obtained:

$$\langle n \rangle = \langle n_0 \rangle \exp[(a - b)t] + \frac{a}{a - b} \{\exp[(a - b)t] - 1\} \tag{5.30}$$

$$\sigma^2 = \langle n^2 \rangle - \langle n \rangle^2$$

$$= \exp[(a - b)t]\langle n_0 \rangle + \frac{a}{a - b} \{\exp[(a - b)t] - 1\}$$

$$+ \frac{2a}{a - b} \{\exp[(a - b)t] - 1\} \exp[(a - b)t]\langle n_0 \rangle \tag{5.31}$$

$$+ \left[\frac{a}{a - b}\right]^2 \{\exp[(a - b)t] - 1\}^2$$

$$+ \exp[2(a - b)t][\langle n_0^2 \rangle - \langle n_0 \rangle^2 - \langle n_0 \rangle]$$

5.4.2 Noise of TWA LD Amp

The single-path gain G_s is obtained by $G_s = \langle n_s \rangle / \langle n_0 \rangle$ from (5.30), where $\langle n_s \rangle$ is the average output photon number relating to the initial value $\langle n_0 \rangle$. At the output, $t = L/V_g$ holds, where L is the length of an amplifier and V_g is the group velocity of light. The result is

$$G_s = \exp[(a - b)(L/V_g)] \tag{5.32}$$

Equations (5.30) and (5.31) are rewritten by using the single-path gain G_s [46]:

$$\langle n_{out} \rangle = G_s \langle n_{in} \rangle + (G_s - 1)n_{sp} m_t \Delta f \tag{5.33}$$

$$\sigma_{out}^2 = \langle n_{out}^2 \rangle - \langle n_{out} \rangle^2$$

$$= G_s \langle n_{in} \rangle + (G_s - 1)n_{sp} m_t \Delta f + 2G_s(G_s - 1)n_{sp}\langle n_{in} \rangle \tag{5.34}$$

$$+ (G_s - 1)^2 n_{sp}^2 m_t \Delta f + G_s^2(\langle n_{in}^2 \rangle - \langle n_{in} \rangle^2 - \langle n_{in} \rangle)$$

$$n_{sp} = a/(a - b) \tag{5.35}$$

where n_{sp} is the population inversion parameter of the amplifying medium. $\langle n_{in} \rangle$, $\langle n_{out} \rangle$, m_t, and Δf are the average input photon number, the average output photon number, the transverse mode number of spontaneous emission light, and the optical bandwidth, respectively. For the single transverse mode, $m_t = 1$. Since (5.30) and (5.31) correspond to the values for a unit optical frequency, Δf is introduced by considering the optical bandwidth for spontaneous emission. The first term of (5.33) represents the amplified signal light and the second term represents the amplified spontaneous emission light. The first and second terms of (5.34) correspond to the shot noises of the amplified signal light and the amplified spontaneous emission light, respectively. The third and fourth terms correspond to the beat noises of the signal-spontaneous emission and the spontaneous emission-spontaneous emission, respectively. The final term represents the signal excess noise and the following equations hold [44]:

$$\langle n_{in}^2 \rangle - \langle n_{in} \rangle^2 - \langle n_{in} \rangle = 0 \quad \text{(for completely coherent light)} \quad (5.36)$$
$$= \langle n_{in} \rangle^2 \quad \text{(for completely incoherent light)}$$

Noise figure F is commonly used for evaluating both electrical amplifiers and optical amplifiers. F is calculated from input S/N and output S/N. The output $(S/N)_{out}$ is obtained by using (5.34) and (5.36), assuming the input light to be a completely coherent light:

$$(S/N)_{out} = \frac{e^2(G_s\langle n_{in}\rangle)^2}{2e^2[G_s\langle n_{in}\rangle + (G_s - 1)n_{sp}m_t\Delta f + 2G_s(G_s - 1)n_{sp}\langle n_{in}\rangle + (G_s - 1)^2 n_{sp}^2 m_t \Delta f]B}$$
(5.37)

where B is the bandwidth of the receiver. The input $(S/N)_{in}$ is obtained by using (5.33):

$$(S/N)_{in} = \frac{e^2\langle n_{in}\rangle^2}{2e^2\langle n_{in}\rangle B} \quad (5.38)$$

By using (5.37) and (5.38), the noise figure F is

$$F = \frac{(S/N)_{in}}{(S/N)_{out}}$$
$$= \frac{1}{G_s} + \frac{G_s - 1}{G_s^2 \langle n_{in}\rangle} n_{sp} m_t \Delta f + 2\left(\frac{G_s - 1}{G_s}\right) n_{sp} + \frac{(G_s - 1)^2 n_{sp}^2 m_t \Delta f}{G_s^2 \langle n_{in}\rangle} \quad (5.39)$$

When the gain G_s is large ($G_s \gg 1$) and $m_t = 1$, F is approximately expressed as

$$F \approx 2n_{sp} + \frac{n_{sp}^2 \Delta f}{\langle n_{in}\rangle} \quad (5.40)$$

For the case that the input signal $\langle n_{in} \rangle$ is large and Δf is small enough, F is

$$F \approx 2 n_{sp} \tag{5.41}$$

For the ideal case, $n_{sp} = 1$ and then $F = 2$ (3 dB). The ideal optical amplifier has a 3-dB noise figure.

The schematic amplifier noise for the general case (TWA LD amp, FP-LD amp, and EDFA) is shown in Figure 5.15. An optical amplifier with gain G amplifies the input signal light and also adds the ASE light to the amplified signal. Not only the shot noises, but the beat noises are also the output light noises. By using the narrowband optical filter, the beat noise of the spontaneous-spontaneous emission can be largely filtered out. The major noise source of the filtered light is the signal-spontaneous emission beat noise.

5.4.3 Noise of FP-LD Amp

The noise of the FP-LD amp is obtained by taking into consideration the FP cavity effect, and the results are [45,46]

$$\langle n_{out} \rangle = G\langle n_{in} \rangle + (G - 1)n_{sp} m_t \Delta f_1 \tag{5.42}$$

$$\begin{aligned}\sigma_{out}^2 &= \langle n_{out}^2 \rangle - \langle n_{out} \rangle^2 \\ &= G\langle n_{in} \rangle + (G-1)n_{sp}m_t\Delta f_1 + 2G(G-1)n_{sp}\chi\langle n_{in}\rangle \\ &\quad + (G-1)^2 n_{sp}^2 m_t \Delta f_2 + G^2(\langle n_{in}^2\rangle - \langle n_{in}\rangle^2 - \langle n_{in}\rangle)\end{aligned} \tag{5.43}$$

$$\chi = \frac{(1 + R_1 G_s)(1 - R_2)(G_s - 1)}{(1 - R_1)(1 - R_2)G_s - (1 - \sqrt{R_1 R_2}\, G_s)^2} \tag{5.44}$$

where G and χ are the gain and the excess noise factor of signal-spontaneous emission beat noise due to the FP cavity, respectively. The gain G is expressed by (5.9). Δf_1 and Δf_2 are the equivalent noise bandwidths for the spontaneous emission shot noise and for the beat noise between spontaneous emission lights, respectively.

Noise figure F is

$$F = \frac{1}{G} + \frac{G-1}{G^2\langle n_{in}\rangle} n_{sp} m_t \Delta f_1 + 2\left(\frac{G-1}{G}\right) n_{sp}\chi + \frac{(G-1)^2 n_{sp}^2 m_t \Delta f_2}{G^2 \langle n_{in}\rangle} \tag{5.45}$$

When the gain G is sufficiently large ($G \gg 1$), the input signal $\langle n_{in}\rangle$ is large, and Δf_2 is small enough, then F is

$$F \approx 2 n_{sp} \chi \tag{5.46}$$

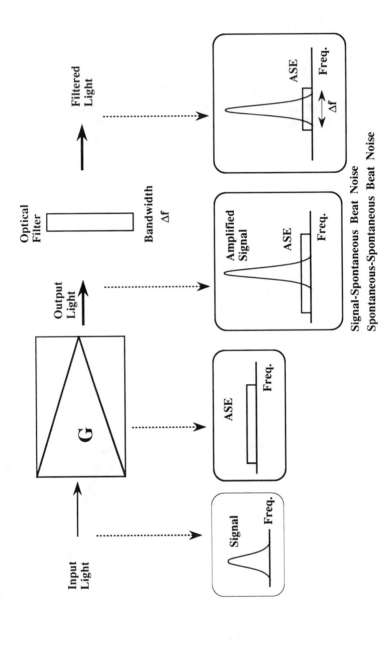

Figure 5.15 Noise of optical amp.

The noise figure F of the FP-LD amp is larger by a factor χ than that of the TWA LD amp when the input signal $\langle n_{in} \rangle$ is large.

5.4.4 Noise of EDFA

The EDFA can be considered one of the TWAs. Therefore, the equations derived for a TWA LD amp are applicable. By considering the two polarization modes for ASE in the EDFA, $m_t = 2$ is used for (5.33) and (5.34). They are

$$\langle n_{out} \rangle = G_s \langle n_{in} \rangle + 2(G_s - 1) n_{sp} \Delta f \tag{5.47}$$

$$\begin{aligned}\sigma_{out}^2 &= \langle n_{out}^2 \rangle - \langle n_{out} \rangle^2 \\ &= G_s \langle n_{in} \rangle + 2(G_s - 1) n_{sp} \Delta f + 2 G_s (G_s - 1) n_{sp} \langle n_{in} \rangle \\ &\quad + 2(G_s - 1)^2 n_{sp}^2 \Delta f + G_s^2 (\langle n_{in}^2 \rangle - \langle n_{in} \rangle^2 - \langle n_{in} \rangle) \end{aligned} \tag{5.48}$$

Noise figure F is

$$F = \frac{1}{G_s} + \frac{2 G_s - 1}{G_s^2 \langle n_{in} \rangle} n_{sp} \Delta f + 2 \left(\frac{G_s - 1}{G_s} \right) n_{sp} + \frac{2(G_s - 1)^2 n_{sp}^2 \Delta f}{G_s^2 \langle n_{in} \rangle} \tag{5.49}$$

When the gain G_s is large ($G_s \gg 1$), F is approximately expressed as

$$F \approx 2 n_{sp} + \frac{2 n_{sp}^2 \Delta f}{\langle n_{in} \rangle} \tag{5.50}$$

5.4.5 Comparison of Optical Amplifiers

There are many differences among the TWA LD amp, the FP-LD amp, and the EDFA regarding the configurations, saturation output power, bandwidth, polarization dependence, gain coefficient for pumping power, and coupling loss with a single-mode fiber. Noise figures for these optical amplifiers are different and are listed in Table 5.2 [47–51]. Noise figure F of the TWA LD amp is smaller than that of the FP-LD amp. This is due to the FP cavity effect. In the EDFA, noise figure F is different for the pumping wavelengths. These differences originates in the different values of the population inversion parameter n_{sp}. For the 0.98-μm pumping wavelength, F is about 3 dB, and this is the theoretical limit of an optical amplifier.

Table 5.2
Noise Figures for LD Amp and EDFA

	LD Amp		Doped-Fiber Amp (EDFA)
	TWA	*FPA*	
Noise figure F	5.2 dB	~10 dB	~4 dB (λ = 0.82-μm pump)
			~3 dB (λ = 0.98-μm pump)
			~4 dB (λ = 1.48-μm pump)
References	[47]	[48]	[49–51]

5.5 OPTICAL AMPLIFIER RESPONSE

Responses to modulated signals are very important for optical amplifiers in the case of optical transmission applications. Although no significant distortion for a TWA LD amp is observed in the small-signal case, pattern-dependent gain (pattern effect) has been reported in the saturation region for 2 Gb/s [52] and for 50 Gb/s [53]. This is due to the gain compression of the LD amp with a few hundred picoseconds (e.g., 500 ps) of gain recovery time τ_g. Time slots of T_s = 500 ps and 100 ps correspond to 2 Gb/s and 10 Gb/s, respectively. For the high-speed data near several gigabits per second, gain compression occurs in an LD amp and the amplified signals depend on the signal patterns (pattern effect). Gain recovery time τ_g plays an important role for the pattern effect in digital transmission and for the waveform distortion in analog transmission. This is shown in Figure 5.16. When time slot $T_s \gg \tau_g$, quick gain recovery in a time slot occurs, and a small waveform distortion (at the pulse wavefront in the digital transmission) is made. This distortion has no significant effect on digital transmission and there is no pattern effect. When time slot $T_s \sim \tau_g$, the gain recovery time takes several time slots and pattern effect and large waveform distortion take place. In the case of $T_s \ll \tau_g$, the gain recovery time takes many time slots and the gain at the specific time slot is determined from the long history of signal patterns and the gain results in the average gain (the gain is constant from the practical point of view). Therefore, there is no pattern effect and no waveform distortion. For the EDFA, gain recovery time τ_g is reported to be about 1 ms [54] and 10 ms [53]. Therefore, $T_s \ll \tau_g$ takes place for many transmission applications, and there is no pattern effect and no waveform distortion for the EDFA. No pattern-dependent gain (pattern effect) for the EDFA has been confirmed for 2 Gb/s [54] and for 100 Gb/s [53]. The AM and FM responses of the EDFA were investigated in the frequency range from 130 MHz to 15 GHz, and the EDFA is found to have a flat response [55].

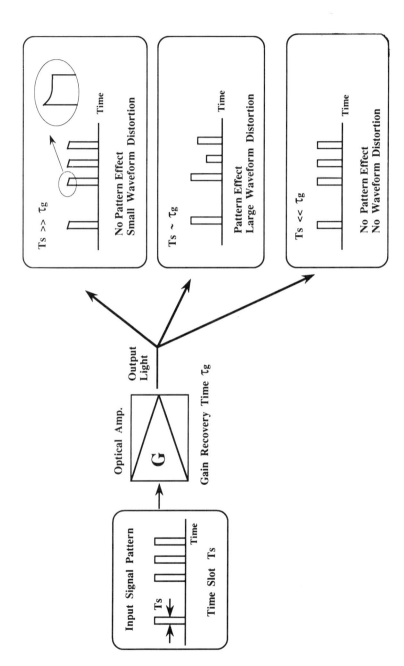

Figure 5.16 Pattern effect and gain recovery time.

5.6 SIMULTANEOUS AMPLIFICATION FOR WDM AND OFDM

Optical amplifiers such as the TWA LD amp and the EDFA have wide bandwidths, with values from 500 to 1,000 GHz, as listed in Table 5.1. This feature is suitable for the simultaneous amplification of several input lights with a different wavelength in WDM or OFDM transmission. The operation of an optical amplifier may sometimes be in the saturation region because of the many input lights. Therefore, the saturation characteristics are important for simultaneous amplification. In general, there are two gain saturation types: homogeneous and inhomogeneous broadening [56], shown in Figure 5.17. In the inhomogeneous case, the gain curve has a hole at the input light wavelength as shown in Figure 5.17, and this effect is called *hole burning*. In the homogeneous case, the shape of the gain curve does not change and only the gain value is affected, as shown in the figure. For simultaneous amplification, it is desirable to have the homogeneous character. It has been reported that both the TWA LD amp and the FP-LD amp can be treated as being homogeneous broadening [57,58]. Although the EDFA is reported not to have the complete homogeneous broadening character [59], it can be treated as being nearly homogeneous broadening for many applications. Therefore, the output power per channel is limited to P_{sat}/N for the case of identical input power per channel in multichannel systems such as WDM or OFDM systems, where P_{sat} is the output saturation power and N is the channel number.

When we use the transmission schemes with a constant optical power such as FSK and PSK, there is no crosstalk between channels for simultaneous amplification, even in the saturation condition. However, crosstalk may occur in the saturation condition for the transmission schemes with nonconstant optical power, such as IM. The mechanism of crosstalk is that the amplifier gain is the function of an input optical power near the saturation, and the output powers of channels are mutually influenced by the input power of channels. There also exists another mechanism for the FP-LD amp, as shown in Figure 5.18. It is a resonance frequency shift (a gain peak frequency shift) [58]. Both gain depletion and resonance frequency shift cause crosstalk in the FP-LD amp. To avoid crosstalk, simultaneous amplification must be made well below the saturation level of an LD amp. For the EDFA, the crosstalk depends on the modulation frequency, and there is no significant crosstalk at frequencies above 100 kHz [60]. This feature of the EDFA is very attractive for multichannel applications.

There are other crosstalk mechanisms for the LD amp in OFDM systems. The nonlinear effect of *nearly degenerate four-wave mixing* (NDFWM) takes place when the frequency interval Δf in OFDM channels is narrow. Crosstalk due to this mechanism occurs when Δf is below several gigahertz, and there is no significant crosstalk when $\Delta f > 2$ to 3 GHz [61,62].

Several characteristics of simultaneous amplification of both the LD amp and the EDFA are summarized in Table 5.3.

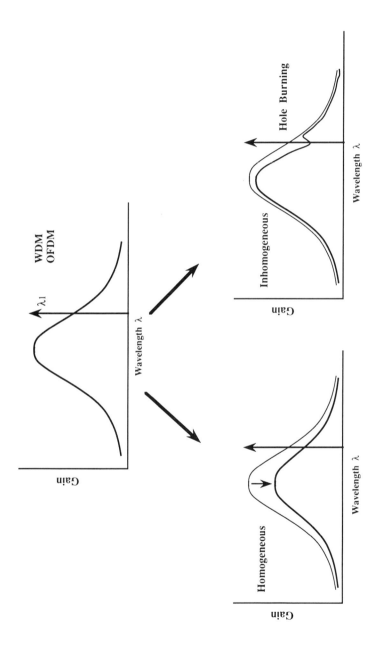

Figure 5.17 Simultaneous optical amplification.

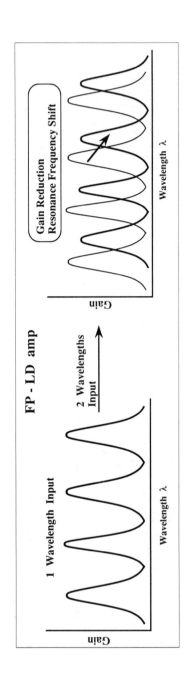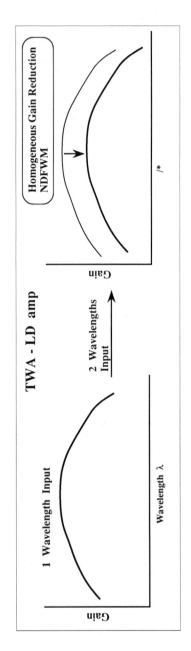

Figure 5.18 Simultaneous amplification of LD amp.

Table 5.3
Simultaneous Amplification Characteristics

	LD Amp (TWA and FPA)	Doped-Fiber Amp (EDFA)
Saturation characteristics	Homogeneous	Nearly homogeneous
Saturation power per channel in N-channel system	$1/N$ (Compared to a single-channel power)	$1/N$ (Compared to a single channel power)
Interchannel crosstalk in OFDM	Crosstalk due to FWM (strong when $\Delta f <$ several GHz)	No crosstalk ($\Delta f > 10$ kHz)
Crosstalk when used in gain saturation state	Crosstalk	No crosstalk ($f > 100$ kHz)

REFERENCES

[1] Saito, T., and T. Mukai, "Recent Progress in Semiconductor Laser Amplifiers," *IEEE J. of Lightwave Technol.*, Vol. LT-6, 1988, p. 1656.
[2] Simon, J. C., "GaInAsP Semiconductor Laser Amplifiers for Single-Mode Fiber Communications," *IEEE J. of Lightwave Technol.*, Vol. LT-5, 1987, p. 1286.
[3] Olsson, N. A., "Lightwave Systems With Optical Amplifiers," *IEEE J. of Lightwave Technol.*, Vol. 7, 1989, p. 1071.
[4] Koester, C. J., and E. Snitzer, "Amplification in a Fiber Laser," *Appl. Opt.*, Vol. 3, 1964, p. 1182.
[5] Hegarty, J., M. M. Broer, B. Golding, J. R. Simpson, and J. B.MacChesney, "Photon Echoes Below 1 K in a Nd^{3+}-Doped Glass Fiber," *Phys. Rev. Lett.*, Vol. 51, 1983, p. 2033.
[6] Poole, S. B., D. N. Payne, R. J. Mears, M. E. Fermann, and R. I. Laming, "Fabrication and Characterization of Low-Loss Optical Fibers Containing Rare-Earth Ions," *IEEE J. of Lightwave Technol.*, Vol. LT-4, 1986, p. 870.
[7] Reekie, L., R. J. Mears, S. B. Poole, and D. N. Payne, "Tunable Single-Mode Fiber Lasers," *IEEE J. of Lightwave Technol.*, Vol. LT-4, 1986, p. 956.
[8] Mears, R. J., L. Reekie, S. B. Poole, and D. N. Payne, "Low-Threshold Tunable CW and Q-Switched Fibre Laser Operating at 1.55 μm," *Electron. Lett.*, Vol. 22, 1986, p. 159.
[9] Mears, R. J., L. Reekie, J. M. Jauncy, and D. N. Payne, "Low Noise Erbium-Doped Fibre Amplifier Operating at 1.54 μm," *Electron. Lett.*, Vol. 23, 1987, p. 1026.
[10] Desurvire, E., J. R. Simpson, and P. C. Becker, "High-Gain Erbium-Doped Travelling-Wave Fiber Amplifier," *Opt. Lett.*, Vol. 12, 1987, p. 888.
[11] Urquhart, P., "Review of Rare Earth Doped Fiber Lasers and Amplifiers," *IEE Proc.*, Vol. 135, 1988, p. 385.
[12] Nakazawa, M., Y. Kimura, and K. Suzuki, "Efficient Er^{3+}-Doped Optical Fiber Amplifier Pumped by a 1.48 μm InGaP Laser Diode," *Appl. Phys. Lett.*, Vol. 54, 1989, p. 295.
[13] Miyajima, Y., T. Komukai, and T. Sugawa, "1.31–1.36 μm Optical Amplification in Nd^{3+}-Doped Fluorozirconate Fibre," *Electron. Lett.*, Vol. 26, 1990, p. 194.

[14] Pedersen, J. E., and M. Brierley, "High Saturation Output Power From a Neodymium-Doped Fluoride Fibre Amplifier Operating in the 1300 nm Telecommunications Window," *Electron. Lett.*, Vol. 26, 1990, p. 819.
[15] Gandy, H. W., R. J. Ginther, and J. F. Weller, "Stimulated Emission of Tm^{3+} Radiation in Silicate Glass," *J. Appl. Phys.*, Vol. 38, 1967, p. 3030.
[16] Reisfeld, R., and Y. Eckstein, "Dependence of Spontaneous Emission and Nonradiative Relaxation of Tm^{3+} and Er^{3+} on Glass Host and Temperature," *J. Chem. Phys.*, Vol. 63, 1975, p. 4001.
[17] Stolen, R. H., "Nonlinear Properties of Optical Fibers," in *Optical Fiber Telecommunications*, S. E. Miller and A. G. Chynoweth, eds., Academic Press, 1979.
[18] Ikeda, M., "Stimulated Raman Amplification Characteristics in Long Span Single-Mode Silica Fibers," *Optics. Comm.*, Vol. 39, 1981, p. 148.
[19] Atkins, C. G., D. Coter, D. W. Smith, and R. Wyatt, "Application of Brillouin Amplification in Coherent Optical Transmission," *Electron. Lett.*, Vol. 22, 1986, p. 556.
[20] Kimura, Y., K. Suzuki, and M. Nakazawa, "46.5 dB Gain in Er^{3+}-Doped Fiber Amplifier Pumped by 1.48 μm GaInAsP Laser Diodes," *Electron. Lett.*, Vol. 25, 1989, p. 1657.
[21] Desurvire, E., C. R. Giles, J. R. Simpson, and J. I. Zyskind, "Efficient Erbium-Doped Fiber Amplifier at a 1.53 μm Wavelength With a High Output Saturation Power," *Opt. Lett.*, Vol. 14, 1989, p. 1266.
[22] Cha, I., M. Kitamura, and I. Mito, "1.5 μm Band Travelling-Wave Semiconductor Optical Amplifiers With Window Facet Structure," *Electron. Lett.*, Vol. 25, 1989, p. 242.
[23] Olsson, N. A., R. F. Kazarinov, W. A. Nordland, C. H. Henry, M. G. Oberg, H. G. White, P. A. Garbinski, and A. Savage, "Polarization-Independent Optical Amplifier With Buried Facets," *Electron. Lett.*, Vol. 25, 1989, p. 1241.
[24] Cha, I., M. Kitamura, H. Honmou, and I. Mito, "1.5 μm Band Travelling-Wave Semiconductor Optical Amplifiers With Window Facet Structure," *Electron. Lett.*, Vol. 25, 1989, p. 1241.
[25] K., Magari, M. Okamoto, H. Yasaka, K. Sato, Y. Noguchi, and O. Mikami, "Polarization Insensitive Traveling Wave Type Amplifier Using Strained Multiple Quantum Well Structure," *IEEE Photon. Lett.*, Vol. 2, 1990, p. 556.
[26] Yamamoto, Y., "Characteristics of AlGaAs Fabry-Perot Cavity Type Laser Amplifier," *IEEE J. of Quantum Electron.*, Vol. QE-16, 1980, p. 918.
[27] Yamamoto, Y., "AM and FM Quantum Noise in Semiconductor Lasers—Part 1: Theoretical Analysis," *IEEE J. of Quantum Electron.*, Vol. QE-19, 1983, p. 34.
[28] Saito, T., and T. Mikai, "1.5 μm GaInAsP Traveling Wave Semiconductor Laser Amplifier," *IEEE J. of Quantum Electron.*, Vol. QE-23, 1987, p. 1010.
[29] Kashima, N., "Analysis of Laser Diode as Transmitter and Preamplifiers in Time Compression Multiplexing Systems," IEEE J. Lightwave Technol., Vol. 10, 1992, p. 323.
[30] Ikagami, T., "Reflectivity of Mode at Facet and Oscillation Mode in Double Heterostructure Injection Lasers," *IEEE J. of Quantum Electron.*, Vol. QE-8, 1972, p. 470.
[31] Reinhalt, F. K., I. Hayashi, and M. B. Panish, "Mode Reflectivity and Waveguide Properties of Double Heterostructure Injection Lasers," *J. Appl. Phys.*, Vol. 42, 1971, p. 4466.
[32] Krupka, D. C., "Selection of Mode Perpendicular to the Junction Plane in GaAs Large-Cavity Double-Heterostructure Lasers," *IEEE J. Quantum Electron.*, Vol. QE-11, 1975, p. 390.
[33] Mukai, T., and Y. Yamamoto, "Gain, Frequency Bandwidth and Saturation Output Power of AlGaAs DH Laser Amplifiers," *IEEE J. of Quantum Electron.*, Vol. QE-17, 1981, p. 1028.
[34] Okamoto, M., K. Sato, H. Mawatari, F. Kano, K. Magari, Y. Kodo, and Y. Itaya, "TM Mode Gain Enhancement in GaInAs-InP Lasers With Tensile Strained-Layer Superlattice," *IEEE J. of Quantum Electron.*, Vol. 27, 1991, p. 1463.
[35] Grosskopf, G., R. Ludwig, R. G. Waarts, and H. G. Weber, "Optical Amplifier Configurations With Low Polarization Sensitivity," *Electron. Lett.*, Vol. 23, 1987, p. 1387.
[36] Olsson, N. A., "Polarization-Independent Configuration Optical Amplifier," *Electron. Lett.*, Vol. 24, 1988, p. 1075.

[37] Eisenstein, G., U. Koren, G. Raybon, T. L. Koch, J. M. Weisenfield, M. Wegener, R. S. Tucker, and B. I. Miller, "Large- and Small-Gain Characteristics of 1.5 μm Multiple Quantum Well Optical Amplifier," *Appl. Phys. Lett.*, Vol. 56, 1990, p. 1201.
[38] Nakazawa, M., Y. Kimura, and K. Suzuki, "Efficient E_r^{3+}-Doped Optical Fiber Amplifier Pumped by a 1.48 μm InGaAsP Laser Diode," *Appl. Phys. Lett.*, Vol. 54, 1989, p. 295.
[39] Suzuki, K., Y. Kimura, and M. Nakazawa, "High Gain E_r^{3+}-Doped Optical Fiber Amplifier Pumped at 800 nm Band," *Electron. Lett.*, Vol. 26, 1990, p. 948.
[40] Vodhanel, R. S., R. I. Laming, V. Shah, L. Curtis, D. P. Bour, W. L. Barnes, J. D. Minelly, E. J. Tarbox, and F. J. Favire, "Highly Efficient 978 nm Diode-Pumped Erbium-Doped Fibre Amplifier With 24 dB Gain," *Electron. Lett.*, Vol. 25, 1989, p. 1386.
[41] Atkins, C. G., J. F. Massicott, J. R. Armitage, R. Wyatt, B. J. Ainslie, and S. P. Caig-Ryan, "High-Gain, Broad Spectral Bandwidth Erbium-Doped Fibre Amplifier Pumped Near 1.5 μm," *Electron. Lett.*, Vol. 25, 1989, p. 910.
[42] Simoda, K., H. Takahashi, and C. H. Townes," Fluctuations in Amplification of Quanta With Application to Maser Amplifiers," *J. Phys. Soc. Japan*, Vol. 12, 1957, p. 686.
[43] Shimoda, K., T. Yajima, Y. Ueda, T. Shimizu, and T. Kasuya, "Quantum Electronics," 2nd ed., Syokabo, 1972 (in Japanese).
[44] Loudon, R., *The Quantum Theory of Light*, Oxford University Press, 1973.
[45] Yamamoto, Y., "Noise and Error Performance of Semiconductor Laser Amplifier in PCM-IM Optical Transmission Systems," *IEEE J. Quantum Electron.*, QE-16, 1980, p. 1028.
[46] Mukai, T., Y. Yamamoto, and T. Kimura, "S/N and Error Rate Performance in AlGaAs Semiconductor Laser Preamplifier and Liner Repeater Systems," *IEEE J. Quantum Electron.*, QE-18, 1982, p. 1560.
[47] Mukai, T., and T. Saito, "5.2 dB Noise Figure in a 1.5 μm InGaAsP Travelling Wave Laser Amplifier," *Electron. Lett.*, Vol. 23, 1987, p. 216.
[48] Mukai, T., T. Saito, and O. Mikami, "1.5 μm InGaAsP Fabry-Perot Cavity Type Laser Amplifiers," *IECE of Japan*, Vol. J69-C, 1986, p. 421 (in Japanese).
[49] Kimura, Y., K. Suzuki, and M. Nakazawa, "Noise Figure Characteristics of E_r^{3+}-Doped Fiber Amplifier Pumped in 0.8 μm Band," *Electron. Lett.*, Vol. 27, 1991, p. 146.
[50] Olshansky, R., "Noise Figure for Erbium-Doped Optical Fiber Amplifiers," *Electron. Lett.*, Vol. 24, 1988, p. 1363.
[51] Desurvire, E., "Analysis of Noise Figure Spectral Distribution in Erbium Doped Fiber Amplifiers Pumped Near 980 and 1480 nm," *Appl. Opt.*, Vol. 29, 1990, p. 3118.
[52] Marshall, I. W., D. M. Spirit, and M. J. O'Mahony, "Picosecond Pulse Response of a Travelling-Wave Semiconductor Laser Amplifier," *Electron. Lett.*, Vol. 23, 1987, p. 818.
[53] Izadpanah, H., D. Chen, Chinlon Lin, M. A. Saifi, W. I. Way, A. YI-Yan, and J. L.Gimlett, "Distortion-Free Amplification of High-Speed Test Patterns up to 100 Gb/s With Erbium-Doped Fiber Amplifiers," *Electron. Lett.*, Vol. 27, 1991, p. 196.
[54] Desurvire, E., and C. R. Giles, "Saturation-Induced Crosstalk in High-Speed Erbium-Doped Fiber Amplifiers at λ = 1.53 μm," *Proc. of Optical Fiber Communication Conference (OFC'89)*, 1989, TUG7.
[55] Laming, R. I., and R. S. Vodanel, "0.1–15 GHz AM and FM Response of Erbium-Doped Fiber Amplifier," *Electron. Lett.*, Vol. 25, 1989, p. 1129.
[56] Yariv, A., *Quantum Electronics*, John Wiley & Sons, 2nd ed., 1975.
[57] Mukai, T., K. Inoue, and T. Saitoh, "Homogeneous Gain Saturation in 1.5 μm InGaAsP Travelling-Wave Semiconductor Laser Amplifiers," *Appl. Phys. Lett.*, Vol. 51, 1987, p. 381.
[58] Jopson, R. M., K. L. Hall, G. Eisenstein, G. Raybon, and M. S. Whalen, "Observation of Two-Color Gain Saturation in an Optical Amplifier," *Electron. Lett.*, Vol. 23, 1987, p. 510.
[59] Inoue, K., H. Toba, N. Shibata, K. Iwatuki, A. Takada, and M. Shimizu, "Mutual Signal Gain Saturation in Er^{3+}-Doped Fibre Amplifier Around 1.54 μm Wavelength," *Electron. Lett.*, Vol. 25, 1989, p. 594.

[60] Pettitt, M. J., A. Hadjifotiou, and R. A. Baker, "Crosstalk in Erbium Doped Fibre Amplifiers," *Electron. Lett.*, Vol. 25, 1989, p. 416.

[61] Inoue, K., T. Mukai, and T. Saitoh, "Nearly Degenerate Four-Wave Mixing in a Travelling-Wave Semiconductor Laser Amplifier," *Appl. Phys. Lett.*, Vol. 51, 1987, p. 1051.

[62] Inoue, K., "Observation of Crosstalk Due to Four-Wave Mixing in a Laser Amplifier for FDM Transmission," *Electron. Lett.*, Vol. 23, 1987, p. 1293.

PART II
System Examples and Optical Devices

Chapter 6
Bidirectional Systems Using TCM

TCM is attractive for low-bit-rate transmission systems, because one fiber is required for bidirectional transmissions and there is no problem of optical reflection. In this chapter, bidirectional transmission systems and several approaches for the economical transceivers possibly used in these systems are explained.

6.1 TCM SYSTEM CONFIGURATION AND EXAMPLES

TCM is briefly explained in Chapter 1 as a multiplexing method for bidirectional transmission. TCM transmission systems are shown in detail in Figure 6.1. In this figure, the configuration and timing chart are shown. In the timing chart, T_D, T_G, and T_{inf} are the delay time, the guard time, and the information time, respectively. These are already explained in Chapter 1. When one transmitter (TX) is in the transmission mode, the other receiver (RX) is in the receiving mode (ping-pong transmission). An example of a frame used in these systems is also shown. In this example, preamble bits (PR), frame synchronous bits (FR), and overhead bits (OH) for housekeeping are inserted as a header of information bits, and cyclic redundancy check (CRC) bits are located at the end of the information bits. The preamble bits carry no information and are only used for the stable performance of a receiver. For a TCM receiver, the receiver acquisition time must be within the time period of the preamble bits.

In TCM, a light source, such as a laser diode, emits about half of the burst cycle T_B. Therefore, there is a possibility of economic transceiver configuration, which is explained later in this chapter. Since signals are transmitted in bursts, reflection light has no effect. The disadvantages of TCM are the increase of line bit rate and transmission delay.

Applications of TCM transmission in subscriber loops are shown in Figure 6.2. TCM transmission is applied to an SS system [1] and a PDS system [2–4]. In the SS system, upstream and downstream signals are transmitted alternately. In the PDS

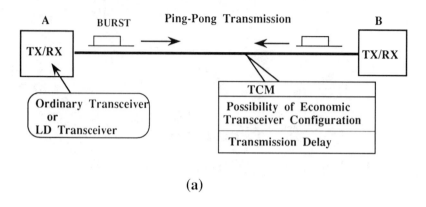

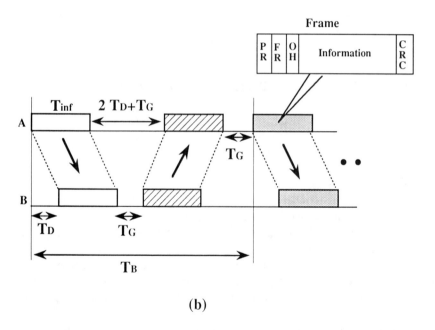

Figure 6.1 TCM transmission systems: (a) configuration; (b) timing and frame.

or generally passive multiple-star systems, upstream and downstream signals are also transmitted alterntively; however, each upstream signals must be controlled to avoid collisions among upstream signals. To avoid collisions, TDMA was proposed [5]. When using TDMA and TCM schemes, each subscriber receives the same down-

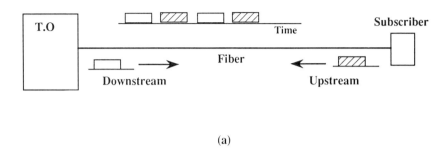

(a)

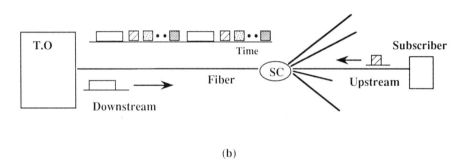

(b)

Figure 6.2 Applications of TCM transmission: (a) SS system; (b) PDS system.

stream signals from the TO and picks up the necessary signals from the downstream signals in the receiving period of TCM. In the sending period of TCM, each subscriber sends upstream signals at the time position determined by TDMA. The TO receives these sequential upstream signals from subscribers.

6.2 LASER DIODE AS A PHOTODETECTOR IN TCM

6.2.1 LD Transceiver in TCM

In TCM systems, laser diodes emit for a period of about a half burst cycle. It is known, but not well known, that a laser diode acts as a photodiode. It is possible to use the same laser diode as both a light emitting device and a photodetecting device in TCM systems. *Laser transceivers* (LD transceivers), in which lasers are used both as light sources and as photodiodes, have been investigated for the low-cost transceivers [6–11,4].

In Figure 6.3, laser transceivers and ordinary transceivers in a TCM transmission system are compared. In the experimental systems in [1], ordinary transceivers were used. In the LD transceiver, optical couplers are removed and a simple configuration is possible.

The configuration of the LD transceiver is shown in Figure 6.4. An electric signal for transmitting is processed in digital processing circuits. In this process, the original signals are converted in the line code and in the TCM burst format. The housekeeping bits are also added. A laser diode is directly modulated by these processed signals and the digitally modulated optical signals are transmitted through a single-mode fiber. In the receiving mode, the laser diode acts as a photodiode and the received electric current is preamplified and amplified with AGC. Timing is extracted and burst signals are digitally determined by the decision circuit. The received TCM burst signals are converted to the original continuous signals by digital processing circuits. The analog process (which means the modulation, amplification, etc.) is controlled by the TCM controller (TCM CONT) based on the information from digital processing circuits.

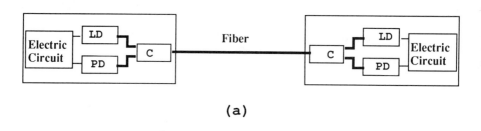

(a)

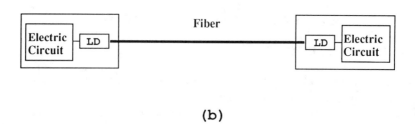

(b)

Figure 6.3 Transceivers in a TCM transmission system: (a) ordinary transceiver; (b) laser transceiver (LD transceiver) (© 1992 *Microwave J.*).

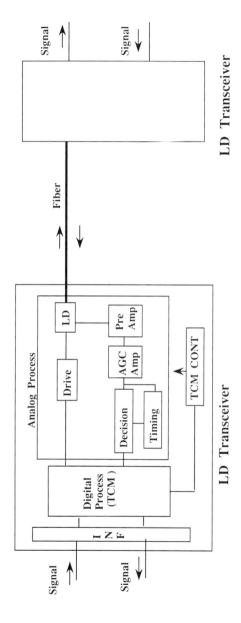

Figure 6.4 Configuration of laser transceiver (© 1992 *Microwave J.*).

6.2.2 Photodetection Properties of a Laser Diode

The basic photodetection properties of a laser diode, such as responsivity and capacitance, must be clarified for the design of a receiver, and they have been investigated in [12,13,6–8]. Polarization dependence and dark-current characteristics have been clarified for a practical receiver design [7,8]. The relationship between the responsivity and the output light power in modules was also clarified [7].

Net Responsivity and Linearity of Responsivity

Responsivity is most important for a laser transceiver. Responsivity R is expressed as $R = \eta R_0$, where η is the coupling efficiency of a light from a fiber with a laser diode, and R_0 is the net responsivity. Using the reciprocity of the light coupling between a laser diode and a single-mode fiber, R_0 can be measured indirectly. One measured net responsivity R_0 of a 1.3-μm Fabry-Perot laser diode chip (FP-(A)) is shown in Figure 6.5. R_0 is 0.5 [A/W]. That is about 60% to 70% of the responsivity of InGaAs pin diodes. Load resistors of 5 and 50 kΩ were used in this measurement. R_0 shows good linearity for both cases.

Relationship Between Responsivity and Output Power in an LD Module

A laser diode is ordinarily packaged in a module, in which it is connected to a single-mode fiber. The coupling efficiency between a laser diode and a single-mode fiber has a great influence on effective responsivity R. Hereafter, a laser module is discussed instead of a laser diode chip, and responsivity R for a laser module includes this coupling efficiency.

Generally, output power P and responsivity R of a laser module are expressed as follows:

$$P = \eta P_{LD} \tag{6.1}$$

$$R = \eta R_0 \tag{6.2}$$

where η, P_{LD}, and R_0 are the coupling efficiency of light from a single-mode fiber into the laser chip, the output power of the laser chip, and the net responsivity of the laser chip, respectively. The reciprocity of the light coupling between the laser chip and single-mode fiber is assumed. By combining (6.1) and (6.2), the following equation holds:

$$R = R_0 P / P_{LD} \tag{6.3}$$

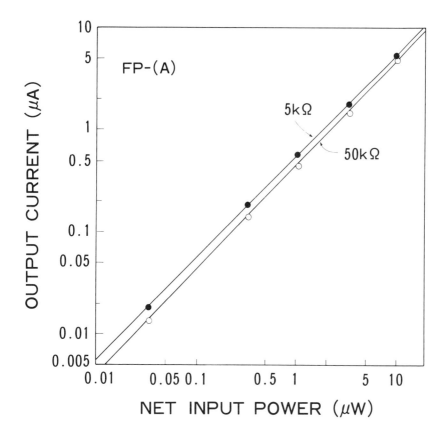

Figure 6.5 Measured output current from a laser as a function of the net input light power (© 1992 IEEE).

When P_{LD} and R_0 are constant for all modules, R is proportional to P. Laser diodes, which are well manufactured, tend to have a constant output power P_{LD} when the injection current ΔI above the threshold current I_{th} and the temperature are constant. From (6.1) and (6.2), the following equation is obtained as follows:

$$R = R_0(dP/dI)/(dP_{LD}/dI) \tag{6.4}$$

where I is the driving current of the laser module. The differentiation of P_{LD} with respect to I, dP_{LD}/dI, is considered to be nearly constant. When R_0 and dP_{LD}/dI are constant for all modules, R is proportional to dP/dI.

Sixty-two commercially available 1.3-μm FP-LD modules were tested, and the measured results, which correspond to (6.3) and (6.4), are shown in Figures 6.6 and 6.7 [7]. When measuring the photodetecting character of a laser module, no bias

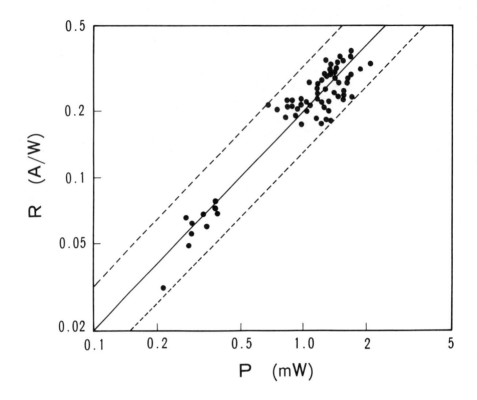

Figure 6.6 The measured responsivity, R, of a laser module as a function of the output power P at $I = I_{th} + 20$ mA (© 1991 IEEE).

voltage is applied (zero bias voltage). The measured R has an approximately linear relationship with P, as shown in Figure 6.6, and is roughly proportional to dP/dI, as shown in Figure 6.7. From these results, the variation in measured R values is considered to originate in the coupling loss variation between a laser chip and a single-mode fiber. It is concluded that the variation of R_0 is small. From Figure 6.6, the following relation is obtained:

$$R = (0.2 \pm 0.1)P \qquad (6.5)$$

where R is in amperes per watt and P is in milliwatts. P is the output power of the module at $I = I_{th} + 20$ mA. Responsivity R can be estimated from the output power of the module using (6.5).

Measured responsivity R for three typical types of LDs are listed in Table 6.1 [8]. Other characteristics are also listed and they are explained later. No difference between an FP-LD and a DFB-LD concerning the R value is recognized from this

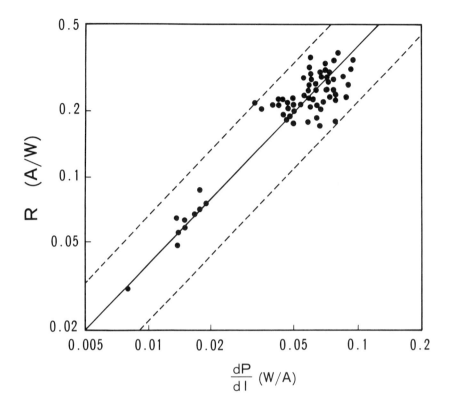

Figure 6.7 Measured responsivity R of a laser module as a function of the differentiation of output power P with respect to current I (© 1991 IEEE).

Table 6.1
Measured Properties of LD Receiver

Name	Type	R at 1.3 μm	C at Zero Bias	Polarization Dependence
FP-(1)	Low-power FP-LD	0.03 [A/W]	80 [PF]	±2%
FP-(2)	Ordinary FP-LD	0.25	50	±3%
DFB-(3)	DFB-LD	0.16	35	±5%

Source: © 1992 IEEE.

result. As can be seen from the above discussion, R of the low-power FP-LD module, which has poor coupling efficiency between a laser and a single-mode fiber, is very small.

Wavelength Dependence of Responsivity

The measured wavelength dependence of responsivity R for 1.3-μm LDs (labeled A through E) is shown in Figure 6.8. The relative responsivity R_r is defined as the ratio of measured responsivity to that at 1.3 μm. From the results, R_r near 1.5 μm is very small for all measured lasers. In a practical system, the wavelength of the

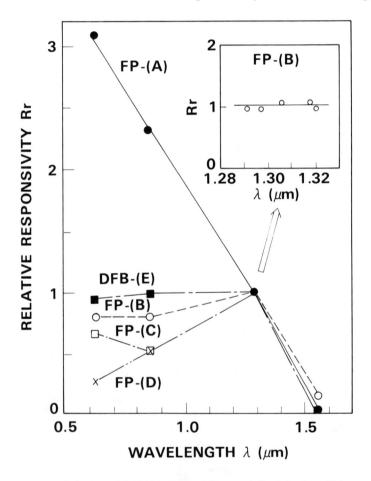

Figure 6.8 Measured relative wavelength dependence of responsivity R for four FP lasers and a DFB laser (© 1992 IEEE).

transmitted light varies around 1.3 μm due to LD manufacturing variations and environmental temperature variations. Thus, it is very important to estimate the variation of R around 1.3 μm. The measured dependence for 34 FP-LD modules is shown in Figure 6.9. While measured R_r varies for individual modules, the average R_r tends to drop slightly at longer wavelengths.

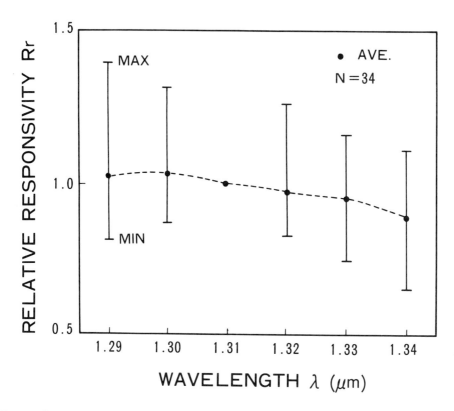

Figure 6.9 Measured relative responsivity R_r as a function of wavelength λ. R_r is defined as the ratio of measured responsivity to that at λ = 1310 nm (© 1991 IEEE).

Influence of Polarization on Responsivity

As discussed in Chapter 5, the gain of an FP-LD amplifier is strongly dependent on polarization. It is important to know the influence of incident light polarization on LD receiver responsivity. The measured polarization dependence is shown in Table 6.1. Responsivity variation due to polarization change ranges from ±2% (relative output is from 0.98 to 1.02) to ±5%. For 55 modules (1.3-μm FP-LD modules), measured responsivity variations due to polarization are ±5% on average, ±1% in minimum and ±16% in maximum.

To determine the origin of the polarization dependence of sensitivity, we make the following calculations. According to [12], responsivity R of a GaAlAs laser can be expressed as

$$R = (e/h\nu)kT\Gamma\alpha_{IB}[1 - \exp(-\alpha L)](1/\alpha) \tag{6.6}$$

where e, $h\nu$, κ, T, Γ, α_{IB}, and L are electron charge, photon energy, coupling efficiency, transmittance coefficient, mode confinement factor, interband absorption, and length of the laser chip, respectively. α is the loss coefficient as defined in (6.7):

$$\alpha = \Gamma\alpha_{IB} + \Gamma\alpha_{FC} + (1 - \Gamma)\alpha_{FCX} + \alpha_s \tag{6.7}$$

where α_{FC}, α_{FCX}, and α_s are the free carrier absorption loss, the free carrier absorption loss outside the active layer, and the scattering loss, respectively. Equation (6.6) was derived assuming that only the active layer contributes the photocurrent. Here, we make the same assumption for an InGaAsP laser. Then (6.6) holds for a 1.3-μm laser. The polarization dependence of R originates from the product term $S (S = T\Gamma)$ and the $[1 - \exp(-\alpha L)](1/\alpha)$ term. The latter term is simply L for small αL. Although αL is not small compared to 1, we ignore the contribution of this term and only consider the term S. The mode confinement factors, Γ_{TE} and Γ_{TM} for the TE and TM modes, respectively, are calculated using a slab model. The transmittance coefficients T_{TE} and T_{TM} for the TE and TM modes are derived from the reflectivity of laser light at a facet and the reciprocity. The reflectivity r for the TE and TM modes is expressed as follows [14,8]:

$$C_m = [\beta_m/4\pi P\omega\mu] \int_{-\infty}^{\infty} |F_{mE}(u)|^2 R_E(u) \, du \quad \text{(TE mode)} \tag{6.8}$$

$$C_m = [\beta_m/4\pi P\omega\varepsilon] \int_{-\infty}^{\infty} |F_{mM}(u)|^2 R_M(u) \, du \quad \text{(TM mode)} \tag{6.9}$$

$$r = |C_m|^2 \tag{6.10}$$

where β_m, P, ω, μ, $F_{mE}(u)$, and $R_E(u)$ are the propagation constant of the mode m, the power carried by the mode, the angular frequency, the permeability, the Fourier coefficient of the electric field at $u (u = \sin \theta)$ for the mode m, and the Fresnel reflection coefficient for the TE mode, respectively. In (6.9), ε, $F_{mM}(u)$, and $R_M(u)$ are the dielectric constant of the material, the Fourier coefficient of the magnetic field, and the Fresnel reflection coefficient for the TM mode, respectively. This method is based on the decomposition of the mode field into plane waves. For TE or TM mode fields, the fields of an active layer and of surrounding layers were calculated by solving the eigenvalue equations for a symmetric three-layer slab waveguide. Numerical results using parameters $\Delta = 9$ and 6% as a function of an active layer thickness 2d are shown in Figure 6.10, where Δ is defined as $(n_1 - n_2)/n_1$ and n_1 and n_2 are the refractive indexes of the active layer and the surrounding layer, respectively. The expression $n_2 = 3.2$ is used in this calculation. The notations in Figure 6.10 are defined as $\Gamma_r = \Gamma_{TM}/\Gamma_{TE}$, $T_r = T_{TM}/T_{TE}$, and $S_r = S_{TM}/S_{TE}$. It is found that $\Gamma_{TE} > \Gamma_{TM}$, while $T_{TE} < T_{TM}$. The output variation of a laser receiver due

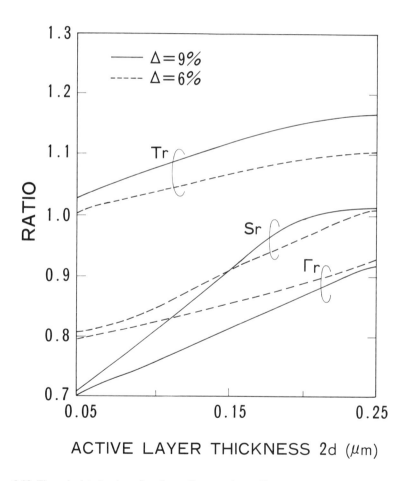

Figure 6.10 The calculated ratios of mode confinement factor Γ and transmittance T and their products S of the TM mode to those of the TE mode as a function of the active layer thickness with a parameter of refractive-index difference Δ (© 1992 IEEE).

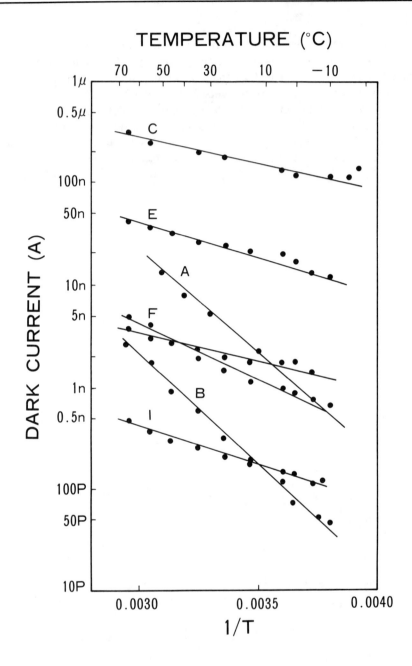

Figure 6.11 Measured dark current as a function of the inverse absolute temperature $1/T$ (© 1992 IEEE).

to the input light polarization is under 1.5 dB for the laser parameters in this figure. Although the precise parameters are not known for lasers in Table 6.1, typical parameters Δ and 2d are around the values of 9% and 0.15 μm for commercial 1.3-μm lasers [15]. If Δ = 9% and 2d = 0.15 μm, then S_r is 0.91. Therefore, the sensitivity of the TE mode is higher than that of the TM mode, and the output variation of a laser receiver due to the input light polarization change is ±4.7%. These results, even though gained with a very simple model, are in good agreement with the experimental results listed in Table 6.1. It may be concluded that the photocurrent is mainly generated in the active layer, and the output variation due to polarization change is under 1 dB for a 1.3-μm InGaAsP laser receiver when operated at zero bias voltage.

Capacitance of a Laser Diode

The capacitance of a laser diode is large when compared to that of a photodiode. Typical capacitance value of a photodiode is 1 pF. As can be seen from the discussion in Chapter 4, the capacitance of a photodetector, in this case a laser diode, affects the sensitivity of an LD transceiver. The measured values of three typical laser modules are listed in Table 6.1. For 62 modules (1.3-μm FP-LD), measured capacitances are 45 pF on average, 19 pF in minimum and 110 pF in maximum. There may be no substantial difference between FP and DFB lasers.

Temperature Dependence of Dark Current

The measured temperature dependence of dark current I_d of seven FP-LD modules (A through F, and I) for small dc bias voltage (0.01V) is shown in Figure 6.11 as a function of the inverse absolute temperature $1/T$. Dark current is expressed by the following equation:

$$I_d = I_0 \exp(-\Delta E/kT) \qquad (6.11)$$

where I_0, ΔE, and k are the constant, the activation energy, and the Boltzmann constant, respectively. Dark currents with forward and reverse biases at small bias voltages are found to be nearly equal (sign is opposite). For smaller bias voltages, the dark current becomes smaller. Therefore, the dark current is estimated to be under 1 μA for bias voltages under 0.01 V at temperatures under 70°C.

6.2.3 Performance of LD Transceiver

Degradation of an LD transceiver compared to a pin-PD receiver is estimated and listed in Table 6.2. The parameters of a pin-PD receiver are assumed to be R = 0.73

Table 6.2
Degradation of LD Receiver Compared to Photodiode Receiver (in dB)

	FP-(1)	FP-(2)	DFB-(3)
Low R	13.9	4.7	6.6
High C	9.5	8.5	7.7
Polarization	0.3	0.3	0.5
Total	23.7	13.5	14.8

Source: © 1992 IEEE.

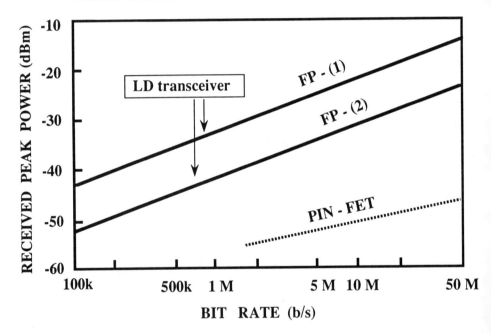

Figure 6.12 Calculated receiver sensitivities (© 1992 IEEE).

[A/W] and $C = 1$ [pF]. The bit rate is assumed to be 1 Mb/s. The main degradation factors are low responsivity R and high capacitance C. The calculated receiver sensitivities of an LD transceiver as a function of bit rate are shown in Figure 6.12. Receiver sensitivities are defined as the received optical power needed for a 10^{-9} error rate. In the same figure, sensitivities of pin-PD receivers are also shown for comparison. The measured LD transceiver sensitivities are shown in Figure 6.13, taken from [6–8,16–18]. Although the laser transceivers have a drawback with regard to the sensitivities when compared to the ordinary transceivers, the simplicity

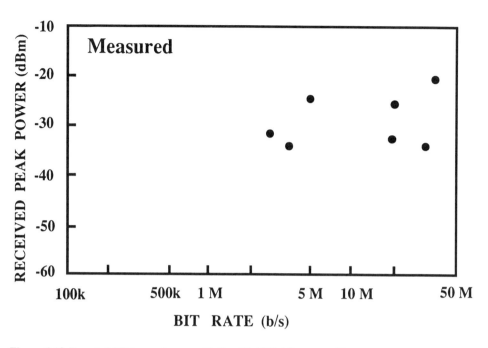

Figure 6.13 Reported LD transceiver sensitivities (© 1992 *Microwave J.*).

in construction and the potential of low-cost transceivers are attractive for optical subscriber loop systems.

6.3 LASER DIODE AS AN ATTENUATOR IN TCM

A wide range of transmission distances from the TO to the subscribers are required to be covered. For reasons of cost efficiency, we need one type of optical receiver that is applicable to both short and long transmission distances. A new method of constructing an optical attenuator for TCM systems is explained here. The transceiver used in this method consists of a laser diode backed by a photodiode. At average power levels, the laser diode is the main photodetector. At higher than average levels, the laser diode acts as an optical attenuator for the photodiode. We call it the *laser attenuator* (LATT). The purpose of this method is to achieve the wide dynamic range of a receiver.

6.3.1 LATT Configuration

To achieve the wide dynamic range of a receiver, one solution is to insert an optical attenuator in front of a high-sensitivity receiver when the received power level exceeds the receiver's saturation level. This solution is not commercially feasible because of its high hardware and maintenance costs.

A new approach to optical attenuation is LATT. The basic TCM system using laser diodes as transceivers is shown in Figure 6.14(a). While this arrangement is very simple, its main drawback is the limited dynamic range. The simplest LATT configuration is shown in Figure 6.14(b). In this figure, LD and PD are a laser diode and a photodiode, respectively. Although LATT does increase the network's dynamic range, some applications require over 40 dB of dynamic range. This is possible with the enhanced LATT configuration shown in Figure 6.14(c). In the enhanced LATT scheme, an additional high-sensitivity PD (PD_2) is coupled to the optical line, and the LATT device is used only as a transmitter and attenuator. At low received power levels, PD_2 is the main photodetector, while PD_1 is used at higher than average power levels.

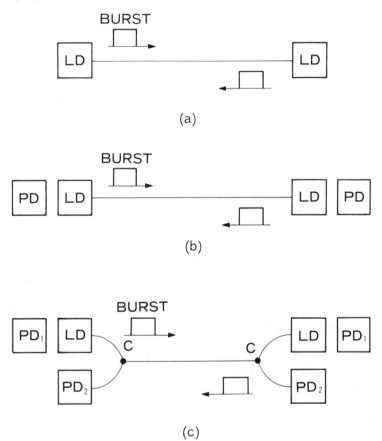

Figure 6.14 TCM systems: (a) TCM system using a laser transceiver; (b) TCM system using a laser transceiver with LATT; (c) TCM system using a coupler with LATT. C represents an optical coupler (© 1991 IEEE).

The receiving and transmitting modes in the basic LATT scheme are shown in Figure 6.15(a,b). In the receiving mode, signals output from the LD and the PD are fed to the separate preamplifiers and then selected according to the received optical power level by the switch (SW). In the transmitting mode, light detected by the PD can be used to control the LD in an automatic power control technique.

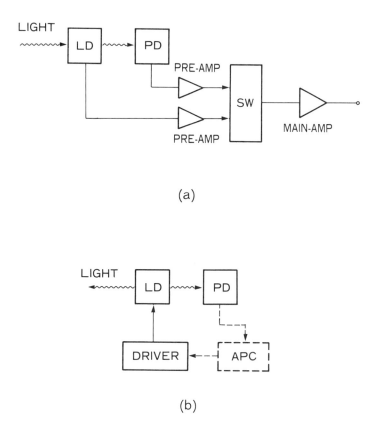

Figure 6.15 Laser transceiver configuration using LATT: (a) receiving mode configuration; (b) emitting mode configuration (© 1991 IEEE).

6.3.2 Basic LATT Properties

Three characteristics of a typical 1.3-μm Fabry-Perot laser module from a commercial source were examined: responsivity, polarization dependence, and wavelength dependence. The module consisted of an InGaAsP laser diode coupled to a germanium photodiode, and it was not modified before being tested.

Responsivity

The responsivities of the LD and the PD are the most important factor in designing an LATT module. The difference of the two responsivities determines the overall performance. The responsivity R and the quantum efficiency of photodetective devices are related as

$$R = \eta e \lambda / (hc) \qquad (6.12)$$

where h, c, e, λ, and η are the Planck's constant, the velocity of light in vacuum, the electronic charge, the wavelength, and the quantum efficiency, respectively. For $\lambda = 1.3$ μm, (6.12) is $R = 1.05 \, \eta$ [A/W]. Responsivity R_{PD} of the PD is expressed as follows:

$$R_{PD} = 1.05 \, \eta = 1.05 \, \exp(-\alpha L) \eta_0 \eta_1 \eta_2 \qquad (6.13)$$

where η_0, η_1, and η_2 are the quantum efficiency of the PD, the coupling efficiency between single-mode fiber and LD, and the coupling efficiency between LD and PD, respectively. The term $\exp(-\alpha L)$ is the loss due to the laser diode at zero bias, where α includes the absorbing and scattering loss in the laser chip and L is the length of the laser chip. Responsivity R_{LD} of the LD and attenuation (ATTEN) of LATT are expressed as follows:

$$R_{LD} = 1.05 \, \eta_1 \eta_3 \qquad (6.14)$$

$$\text{ATTEN} = -10 \log(R_{PD}/R_{LD}) \qquad (6.15)$$

η_3 is the quantum efficiency of the laser in the receiving mode. These equations show that the attenuation in LATT originates not only from laser absorption but also from coupling efficiency. To obtain the order of R and ATTEN, the following values are assumed: $\eta_0 = 0.75$, $\eta_1 = 0.4$, $\eta_2 = 0.2$, and $\exp(-\alpha L) = 0.3$. The calculated R_{PD} is 0.019 [A/W]. If $\eta_1 = 0.4$ and $\eta_3 = 0.5$ are assumed, then $R_{LD} = 0.21$. The attenuation of LATT is estimated to be about 10 dB with the assumed values.

$R_{LD} = 0.29$ [A/W] and $R_{PD} = 0.014$ [A/W] at a wavelength of $\lambda = 1306$ nm were measured for a 1.3-μm FP-LD module, and the attenuation was 13.2 dB. A 1.5-μm FP-LD module was also measured at $\lambda = 1535$ nm wavelength, and the results were $R_{LD} = 0.22$ [A/W], $R_{PD} = 0.014$ [A/W], and ATTEN $=12.0$ dB.

Polarization Dependence

R_{LD} (laser diode responsivity) is influenced by the polarization of the input light, as explained in the previous section. This implies that LATT is also polarization-dependent. The polarization dependence of R_{LD} originates from the product term

$S (S = \Gamma T)$ of a simplified laser model, where T and Γ are the transmittance coefficient at the front laser facet and the mode confinement factor, respectively. T and Γ are different from the TE mode to the TM mode. Based on the same model, the polarization dependence of R_{PD} originates from the product term X, which is defined in (6.16):

$$X = T^2 \Gamma \qquad (6.16)$$

Equation (6.16) indicates that the light passing through a laser chip encounters the front facet (T), an active layer (Γ), and the rear facet (T). An equation for calculating T is

$$T = 1 - |C_m|^2. \qquad (6.17)$$

C_m is calculated by (6.8) for the TE mode and (6.9) for the TM mode.

Numerical calculations based on the derived equations were made using the parameters $\Delta = 9\%$ and 6% as a function of active layer thickness (2d) and are shown in Figure 6.16. Δ is defined as the refractive-index difference of the active layer in a laser chip. Ratios X_r and S_r are defined as $X_r = X_{TM}/X_{TE}$ and $S_r = S_{TM}/S_{TE}$.

Responsivity variation due to the polarization state can be obtained from this figure. For example, when $\Delta = 9\%$ and $2d = 0.17$ μm, $S_r = 0.94$ and $X_r = 1.06$. Therefore, the output variations of an LD receiver and a PD receiver with LATT are ± 3.1 and $\pm 2.9\%$, respectively. The maximum output variation of a PD receiver (with LATT) is $\pm 12.4\%$ for the parameters assumed in the calculations. The polarization dependence of R was measured to be $\pm 3.7\%$ and $\pm 3.2\%$ for the LD receiver and PD receiver (with LATT), respectively, using a light source with a 0.94-degree of polarization.

Wavelength Dependence

The responsivity R_{LD} of a 1.3-μm laser diode drops around 1.5 μm and is nearly constant around 1.3 μm [8,13,19]. Therefore, R_{PD} is considered to be nearly constant around 1.3 μm. It was confirmed experimentally that LATT attenuation is nearly constant for the measured wavelength interval of 2.4 nm around $\lambda = 1.3$ μm [9].

6.3.3 Dynamic Range of Receiver With LATT

The dynamic range of the receiver shown in Figure 6.15(a) is considered. With LATT, the total dynamic range D_T can be increased by the value of ATTEN (the attenuation of LATT). By lowering the sensitivity of the PD receiver (this is possible

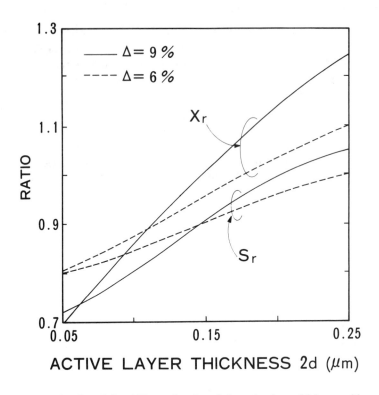

Figure 6.16 Calculated ratios of S and X as a function of the active layer thickness with a parameter of a refractive-index difference Δ (© 1991 IEEE).

by choosing preamplifier design parameters), the PD receiver can receive a high-power light signal. When assuming the same dynamic range for LD and PD receivers, the following estimation holds:

$$D_{LD} + (ATTEN) \leq D_T \leq 2 D_{LD} \tag{6.18}$$

where D_{LD} is the dynamic range of the LD receiver and ATTEN is defined by (6.15). For example, if ATTEN = 15 dB and D_{LD} = 25 dB, then 40 dB $\leq D_T \leq$ 50 dB.

To confirm the dynamic range increase possible with LATT, the results of fundamental transmission experiments are shown in Figure 6.17. The measured error rate (circles) at 3.5 Mb/s and the dynamic range (arrows) are shown in this figure for the LD and the PD. The dynamic range is defined as the optical receiver power interval at an error rate of 10^{-9}. D_{LD} was measured to be about 24 dB. D_{PD} was not measured because the received power could not be higher than +1.1 dBm in this experiment. At +1.1 dBm, signals were received by the PD receiver (with LATT) without error. The total dynamic range D_T was over 37 dB. ATTEN of the LATT

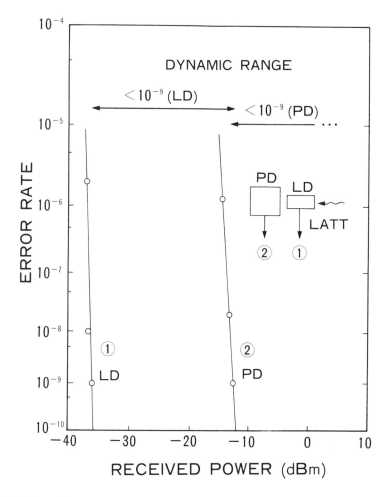

Figure 6.17 Measured error rate as a function of the received power and the measured dynamic range. LD and PD indicate the laser receiver and the PD receiver (with LATT), respectively (© 1991 IEEE).

used (one of the commercial 1.3-μm FP-LD modules) was 13.2 dB. Therefore, the total dynamic range was at least increased by the value of ATTEN. By overlapping the receiving ranges of LD and PD receiving, automatic switching between LD and PD receiving is made according to the input receiving light level. By switching at the interval of the guard time T_G, no signal interruption is made. This was experimentally confirmed.

If LATT is used in the enhanced configuration (Figure 6.14(c)), a wider dynamic range can be realized because the receiver sensitivity of PD_2 can be lower than that of the laser receiver. At 3.5 Mb/s, −50 dBm of receiver sensitivity may be possible, and 0 dBm may be received by using a PD (with LATT). This results in a 50-dB dynamic range.

6.4 LASER DIODE AS A PREAMPLIFIER IN TCM

A novel approach that used a laser preamplifier at a low-bias current in a TCM system is explained here [20]. The low-bias current approach creates wavelength- and polarization-independent amplifiers. The receiver sensitivity of this approach is estimated and compared with a laser transceiver.

6.4.1 Receiver Configuration

The conventional system using LD and PD is shown in Figure 6.18(a). This configuration needs two couplers and the loss of couplers (about 3 dB each) must be considered in the loss budget. The system using LD transceivers is shown in Figure 6.18(b). A novel approach that used a laser preamplifier in a TCM system is shown in Figure 6.18(c). The LD acts as a light emitter in the transmission mode and as a preamplifier in the receiving mode. The PD is used as a photodetecting device for automatic power control of the laser power in the transmission mode and as a signal receiver in the receiving mode.

As explained in the previous chapter, it is known that the laser becomes an FP-type amplifier with sufficient gain. The disadvantages of FP-type amplifiers are (1) large wavelength dependence and (2) large polarization dependence. These dependencies result in the need for polarization and temperature controllers and the selection of lasers with the same lasing wavelength. Another difficulty is that FP lasers have many lasing modes (multimode lasing). These problems are formidable barriers to the construction of economical subscriber systems. The novel approach is based on a laser that acts as a traveling wave amplifier and transmitter. The LD acts as a traveling wave amplifier with high facet reflectivity when operated at low bias levels. Although the gain is small, an LD with low bias levels has none of the disadvantages of FP-type amplifiers. As for the polarization dependence, it is known from Figure 5.8 that an LD with low bias levels has no significant dependence. No significant wavelength dependence was confirmed by measurements [20].

6.4.2 Transmission Experiment Using a Conventional Laser Module

The experimental setup, which is intended to clarify the receiver sensitivity improvement with the low-bias current preamplifier, is shown in Figure 6.19. ATT, AMP,

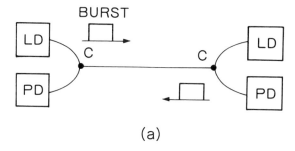

(a)

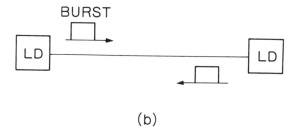

(b)

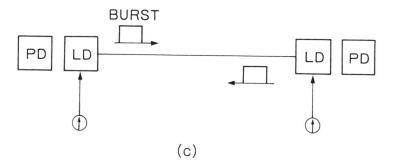

(c)

Figure 6.18 TCM transmission systems: (a) conventional system using LD and PD; (b) system using laser transceiver; (c) system using a laser diode as a preamplifier in the receiving mode (© 1992 IEEE).

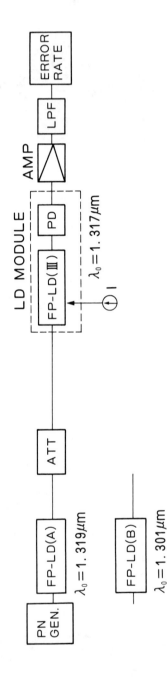

Figure 6.19 Experimental setup for measuring the error rate. Two FP-LDs are used as light sources. FP-LD (III) is used as an optical preamplifier (© 1992 IEEE).

and LPF are an optical attenuator, an electrical amplifier, and a low-pass filter, respectively. Electrical signal patterns generated by a *pseudorandom noise* (PN) generator (PN GEN) were converted to optical signals and were detected. An FP-type LD (named FP-LD (III)) was used as a preamplifier, and the bias current I was set to $I = 0$ or $I = 0.54I_{th}$. The monitor photodiode (Ge-pin) was used as a photodetector. The lasing wavelength λ_0 of this FP-LD is 1317 nm. Two multimode lasers (FP-LD (A) and FP-LD (B)) were used as light sources and were modulated with a pseudorandom pattern ($2^{15} - 1$) at 3.5 Mb/s. The lasing wavelengths of FP-LD (A) and (B) are 1319 nm (near λ_0) and 1301 nm (not near λ_0), respectively. The measured error rates as a function of average power are shown in Figure 6.20. The bias currents

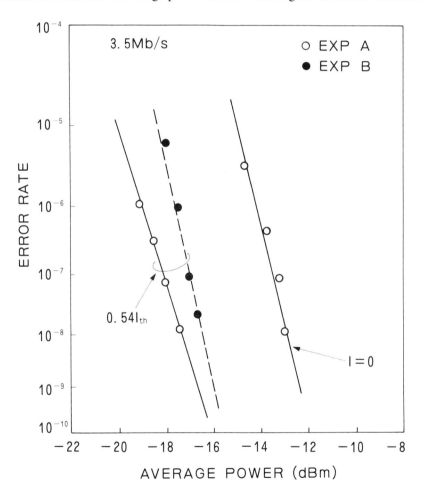

Figure 6.20 Measured error rate as a function of received average power (© 1992 IEEE).

of FP-LD(III) were set at $I = 0$ and $I = 0.54I_{th}$. EXP A and B used FP-LD (A) and (B), respectively. From the results of Figure 6.20, about a 4-dB improvement in receiver sensitivity was obtained at $I = 0.54I_{th}$ when compared to $I = 0$. There is little difference in receiver sensitivity improvement for FP-LD (A) and (B) at $I = 0.54I_{th}$, which means little wavelength dependence.

6.4.3 Receiver Sensitivity

The experiment reveals that receiver sensitivity can be improved by operating a laser preamplifier at low-bias current levels. However, the module used in this experiment was not optimized for the proposed configuration. The monitor PD was not designed for signal detection, and the estimated coupling loss between the LD and the monitor PD is large. This section analyzes an optimized configuration. The receiver sensitivity with an optimum laser preamplifier is estimated by considering noise.

Noise

Laser preamplifier. The noise of a traveling wave laser amplifier is expressed as follows [21,22] (this is discussed in Chapter 5):

$$\sigma^2 = \langle n_{out}^2 \rangle - \langle n_{out} \rangle^2 \qquad (6.19)$$
$$= G_s \langle n_{in} \rangle + (G_s - 1)n_{sp}m_t \Delta f + 2G_s(G_s - 1)n_{sp}\langle n_{in} \rangle + (G_s - 1)^2 n_{sp}^2 m_t \Delta f$$

$$G_s = \exp[(\Gamma g - \alpha)L] \qquad (6.20)$$

$$n_{sp} = \frac{N}{N - N_0} \frac{\Gamma g}{\Gamma g - \alpha} \qquad (6.21)$$

$$g = A_g(N - N_0) \qquad (6.22)$$

where:

$\langle n_{in} \rangle$ = average input photons per second;
$\langle n_{out} \rangle$ = average output photons per second;
σ = output variation of number of photons;
n_{sp} = population inversion parameter;
Δf = optical bandwidth;
m_t = transverse mode number;
Γ = mode confinement factor;
α = absorption coefficient (loss coefficient);

g = gain coefficient;
L = laser cavity length;
A_g = differential gain coefficient ($= dg/dN$);
N = injected carrier density; and
N_0 = carrier density where $g = 0$.

To apply these equations to a traveling wave amplifier with highly reflective facets, (6.19) is modified by considering the facet loss (reflection at facets) and the coupling loss. The following equation is obtained:

$$\sigma^2 = \eta_1\eta_2\eta_3\eta_4\eta_5 G_s\langle n_{in}\rangle + \eta_3\eta_4\eta_5(G_s - 1)n_{sp}m_t\Delta f \\ + 2\eta_1\eta_2(\eta_3\eta_4\eta_5)^2 G_s(G_s - 1)n_{sp}\langle n_{in}\rangle + (\eta_3\eta_4\eta_5)^2(G_s - 1)^2 n_{sp}^2 m_t\Delta f \quad (6.23)$$

where:

η_1 = coupling efficiency between laser diode and single-mode fiber;
η_2 = transmission coefficient ($= 1 - R$) at the input laser facet;
η_3 = transmission coefficient ($= 1 - R$) at the output laser facet;
η_4 = coupling efficiency between laser diode and photodiode; and
η_5 = quantum efficiency of photodiode.

In order to calculate the noise, the following laser parameters are assumed: reflectivity R at the facets $= 0.32$, $\Gamma = 0.36$, $L = 300$ μm, $A_g = 4 \times 10^{-16}$ [cm^2], and $N_0 = 9 \times 10^{17}$ [cm^{-3}]. Strictly speaking, R and Γ are different for TE and TM modes. However, this was ignored in the calculations because there is no significant difference in the calculated results. Parameters η_1, η_2, η_3, η_4, and η_5 are assumed to be 0.5, 0.68, 0.68, 1, and 0.67 (0.67 corresponds to 0.7 [A/W] at 1.3 μm), respectively. The optimized configuration is assumed to be 0 dB ($\eta_4 = 1$) for the coupling of laser diode and photodiode. Bias current $I = 0.54 I_{th}$ is assumed. To determine $m_t\Delta f$, spontaneous emission is assumed to be 1.1×10^{-6} [W] based on the measured value. For calculating g, the measured gain value of 2 at $I/I_{th} = 0.54$ in Figure 5.8 is used with (6.20).

Photodiode. The assumed parameters are 0.7 [A/W] for sensitivity and 1 pF for capacitance C and 10 nA for the dark current.

Electrical preamplifier. The electrical preamplifier generates thermal noise. This thermal noise (circuit noise) is assumed to be 0.24 pA/$\sqrt{\text{Hz}}$ for a good amplifier and 5 pA/$\sqrt{\text{Hz}}$ for a poor amplifier at 1 MHz. Circuit noise is assumed to increase linearly with bit rate B. The electric preamplifier bandwidth is assumed to be $0.7B$.

Calculated Sensitivity With an Optimized Configuration

Receiver sensitivity can be calculated using the assumed noise in the section above. Receiver sensitivity is defined as the input power of a laser module at an error rate of 10^{-9}. Calculated results for the optimized configuration are shown in Figures 6.21 and 6.22. For 1.3-μm InGaAsP lasers, the laser loss parameter $\alpha = 42.8$ cm^{-1} for $\Gamma = 0.36$ can be estimated using the data obtained for 1.3-μm InGaAsP lasers in [23], and nearly equal data are expected from the data given in [24]. For 1.5 μm-InGaAsP lasers, $\alpha = 29$ cm^{-1} for $\Gamma = 0.36$ were reported [22]. In this calculation, $\alpha = 20$ cm^{-1} and 60 cm^{-1} are used as parameters. Figures 6.21 and 6.22 correspond to the use of good and poor electrical preamplifiers, respectively. Receiver sensitivity of the conventional configuration was also calculated for comparison. These are indicated with PD in these figures. The conventional configuration uses two couplers,

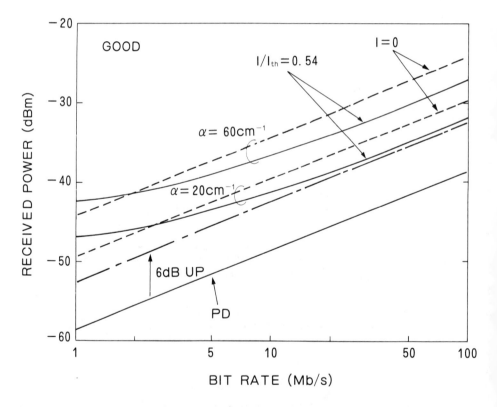

Figure 6.21 Calculated receiver sensitivity when using a good electrical preamplifier as a function of bit rate. PD and 6-dB UP notations in this figure are the results for the conventional configuration. Others are for the proposed configuration with bias current $I = 0$ and $I/I_{th} = 0.54$. Laser absorption coefficient α is used as a parameter (© 1992 IEEE).

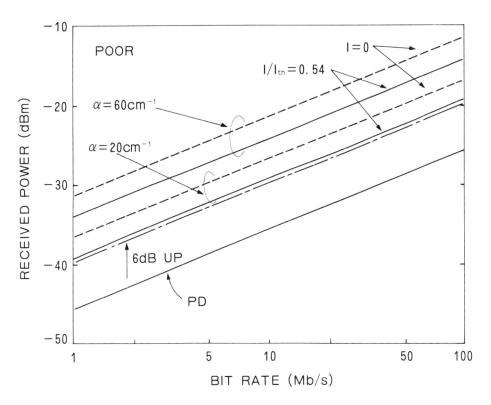

Figure 6.22 Calculated receiver sensitivity when using a poor electrical preamplifier as a function of bit rate. PD and 6-dB UP notations in this figure are the results for the conventional configuration. Others are for the proposed configuration with bias current $I = 0$ and $I/I_{th} = 0.54$. Laser absorption coefficient α is used as a parameter (© 1992 IEEE).

and this results in a 6-dB loss. Therefore, the effective received power is 6 dB higher, assuming the same output transmitter power among these configurations. In these figures, 6-dB UP curves, which correspond to the conventional configuration with two couplers, are also plotted.

From these figures, the following results are obtained for the assumed parameters. In the case of a good electrical preamplifier, the sensitivity with a laser preamplifier at $I/I_{th} = 0.54$ bias is better than that with a zero bias laser for bit rates higher than several megahertz. For bit rates lower than several megahertz, the converse is true. The spontaneous-spontaneous and signal-spontaneous beat noises and spontaneous shot noise are believed to exceed the thermal noise (circuit noise) at low bit rates. Thus, these noises cancel the improvement by amplification at $I/I_{th} = 0.54$ bias. The thermal noise dominates at high bit rates because it is assumed to increase linearly with the bit rate. In this situation, the noise of the amplifier is hidden behind the thermal noise, and the benefit of amplification can be seen. The sensitivities for a = 20 cm^{-1} and 60 cm^{-1} are 1.1 and 5.7 dB poorer, respectively, at 10 Mb/s than that of the conventional configuration with 6-dB loss (coupler losses).

In the case of a poor electrical preamplifier, about a 2.7-dB improvement in receiver sensitivity is expected with laser preamplification at $I/I_{th} = 0.54$ bias when compared with $I = 0$ for all the bit rates considered in Figure 6.22. The thermal noise due to the poor amplifier is the dominant noise, and this results in a sensitivity improvement with amplification. The sensitivity for $\alpha = 20$ cm^{-1} is nearly equal and that for 60 cm^{-1} is 5.5 dB poorer at 10 Mb/s, when compared with the conventional configuration with 6-dB loss.

Comparison With a Laser Transceiver

It is interesting to compare the LD preamplifier configuration with the LD transceiver in terms of receiver sensitivity. The sensitivity of 0.25 [A/W], which includes the fiber coupling loss, is assumed for the LD transceiver. The capacitances C of the laser module are assumed to be 5 pF and 20 pF. The sensitivity of the laser transceiver is assumed to be proportional to $(\sqrt{C})^{-1}$. The calculated sensitivities with the good preamplifier are shown in Table 6.3. The result is that the sensitivity of the LD preamplifier with optimized configuration is similar to that of the LD transceiver.

Table 6.3
Calculated Receiver Sensitivity (in dBm)

Bit Rate (Mb/s)	PD (6 dB up)	LD Preamplifier Configuration		LD Transceiver	
		($\alpha = 20$ cm^{-1})	($\alpha = 60$ cm^{-1})	(20 pF)	(5 pF)
4	−46.4	−44.2	−39.6	−41.4	−44.4
30	−38.0	−37.9	−32.3	−33.0	−36.0
100	−32.5	−32.0	−27.5	−27.5	−30.5

Source: © 1992 IEEE.

Comparison With Experiment

The transmission experiment in Section 6.4.2 corresponds to the poor electrical preamplifier case. The measured improvement is about 4 dB, while the calculated one is about 2.7 dB for the assumed parameters. This discrepancy is due to the fact that the module used in the experiment was not optimized. When $\eta_4 = 0.25$ (6 dB) is taken instead of $\eta_4 = 1$ (0 dB), the calculated sensitivity difference (improvement) between $I = 0$ and $I/I_{th} = 0.54$ for 3.5 Mb/s ($\alpha = 60$ cm^{-1}) is 4.2 dB. This is closer to the measured improvement.

REFERENCES

[1] Jaillard, A., H. Prigent, and Y. Guillausseau, "Experimental Data Link Using Single Mode Fibers and Ping-Pong Transmission," *EFOC LAN '89*, Amsterdam, 1989, p. 66.
[2] Bourgt, F., "Optical Passive Distribution in the Local Network," IEEE Workshop on Passive Optical Networks for the Local Loop, No. 3.5, London, 1990.
[3] Okada, K., and F. Mano, "Passive Double Star System Features," The Third IEEE Workshop on Local Optical Networks, No. 5.1, Tokyo, 1991.
[4] Abiven, J., "TOME: A New Concept for Passive Optical Network," The Third IEEE Workshop on Local Optical Networks, No. 5.2, Tokyo, 1991.
[5] Stern, J. R., J. W. Ballance, D. W. Faulkner, S. Hornung, and D. B. Payne, "Passive Optical Local Network for Telephony Applications and Beyond," *Electron. Lett.*, Vol. 23, 1987, p. 1255.
[6] Kashima, N., "A Study of Time Compression Multiplex Transmission System Using a 1.3 μm Semiconductor Laser as a Transmitter/Receiver," No. B-949 in *Proceedings of the 1990 IEICE*, spring conference, March 1990 (in Japanese).
[7] Kashima, N., "Properties of Commercial 1.3 μm Fabry-Perot Laser Modules in a Time Compression of Multiplexing System," *IEEE J. Lightwave Technol.*, Vol. 9, 1991, p. 918.
[8] Kashima, N., "Time Compression of Multiplex Transmission System Using a 1.3 μm Semiconductor Laser as a Transmitter and a Receiver," *IEEE Trans. Commun.*, Vol. 40, 1992, p. 584.
[9] Kashima, N., "A New Approach to an Optical Attenuator for a Time Compression Multiplex System Using a Laser Diode as Both Transmitter and Receiver," *IEEE J. Lightwave Technol.*, Vol. 9, 1991, p. 987.
[10] Bergmann, E. E., and D. A. Snyder, "Time Compression Techniques for Full-Duplex Communications on One Fiber," *J. Opt. Commun.*, Vol. 12, 1991, p. 91.
[11] Van Der Ziel, J. P., R. M. Mikulyak, and R. A. Logan, "7.5 km Bidirectional Singlemode Optical-Fiber Link Using Dual-Mode InGaAsP/InP 1.3 μm Laser Detectors," *Electron. Lett.*, Vol. 21, 1985, p. 511.
[12] Alping, A., R. Tell, and S. T. Eng, "Photodetection Properties of Semiconductor Laser Diode Detectors," *IEEE J. Lightwave Technol.*, Vol. LT-4, 1986, p. 1162.
[13] Van Der Ziel, J. P., "Characteristics of 1.3 μm InGaAsP Lasers Used as Photodetectors," *IEEE J. Lightwave Technol.*, Vol. 7, 1989, p. 347.
[14] Krupka, D. C., "Selection of Mode Perpendicular to the Junction Plane in GaAs Large-Cavity Double-Heterostructure Lasers," *IEEE J. Quantum Electron.*, Vol. QE-11, 1975, p. 390.
[15] Itaya, Y., Y. Suematu, S. Kitayama, K. Kishina, and S. Arai, "Low Threshold Current Density (100) GaInAsP/InP Double-Heterostructure Lasers for Wavelength 1.3 μm," *Japan. J. Appl. Phys.*, Vol. 18, 1979, p. 1795.
[16] Shuji, M., M. Akita, S. Makino, and T. Kitayama,"Burst Transmission Characteristics of Optical transceiver for Single Star Local Loop," *Proceedings of the 1991 IEICE of Japan*, spring conference, No. B-907, March 1991 (in Japanese).
[17] Ishizaki, H., K. Takeda, I. Fukuzaki, K. Yusa, and K. Nakamura, "A Study on Optical Receiver Using Laser Diode for PDS Fiber-Optic Access Network System," *Proceedings of the 1992 IEICE of Japan*, spring conference, No. B-749, March 1992 (in Japanese).
[18] Faulkner, D. W., A. Hunwicks, D. B. Payne, and J. R. Stern, "Opto-Electric Device Requirements for Local Loop Passive Optical Networks," The Third IEEE Workshop on Local Optical Networks, No. 7.1, Tokyo, 1991.

[19] Nakazawa, M., T. Nakashima, and S. Seikai, "Self-Detecting Optical Time Domain Reflectometer for Single Mode Fibers," *Opt. Lett.*, Vol. 10, 1985, p. 157.
[20] Kashima, N., "Analysis of Laser Diode as Transmitter and Preamplifiers in Time Compression Multiplexing Systems," *IEEE J. Lightwave Technol.*, Vol. 10, 1992, p. 323.
[21] Yamamoto, Y., "Noise and Error Performance of Semiconductor Laser Amplifier in PCM-IM Optical Transmission Systems," *IEEE J. Quantum Electron.*, QE-16, 1980, p. 1028.
[22] Mukai, T., T. Saito, and O. Mikami, "1.5 μm InGaAsP Fabry-Perot Cavity Type Laser Amplifiers," *IECE of Japan*, Vol. J69-C, 1986, p. 421 (in Japanese).
[23] Itaya, Y., Y. Suematu, S. Katayama, K. Kishino, and S. Arai, "Low Threshold Current Density (100) GaInAsP/InP Double-Heterostructure Lasers for Wavelength 1.3 μm," *Japan J. Appl. Phys.*, Vol. 18, 1979, p. 1795.
[24] Naholy, R. E., M. A. Pollock, and J. C. Dewinter, "Temperature Dependence of InGaAsP Double-Heterostructure Laser Characteristics," *Electron. Lett.*, Vol. 15, 1979, p. 695.

Chapter 7
Optical Transmission Systems Using WDM

WDM technologies, which make it possible for several users or channels to share one fiber, will be used for several transmission systems in subscriber loops. In this chapter, upgrading and multichannel systems using WDM are explained. WDM devices used for these systems are discussed from the viewpoint of system design.

7.1 WDM FOR UPGRADING METHOD

7.1.1 Upgrading Methods

Generally, optical trunk systems transmit many subscribers' signals through the multiplexing, and one subscriber's service demand has no significant influence. An optical subscriber loop system is strongly influenced by each subscriber's service demand. Suppose that the initial introduction of an optical transmission system to one subscriber is made by the specific service (here, we call it service A). Later, this subscriber may require other services, such as B, C, and so on. This service upgrade must be made smoothly. Several upgrading methods can be considered, and some of them are shown in Figures 7.1 to 7.3. The upgrade method by electrical digital multiplexing using the STM or ATM is shown in Figure 7.1. Although the number of wavelengths depends on the schemes for bidirectional transmission (TCM, DDM, SDM, etc.), only one or two wavelengths are deployed. The drawbacks of this method are the need for a higher bit rate transmission system and the service interruption. The ATM has the robustness for an increase in services at the cost of quality. Therefore, the increase in services for each subscriber requires higher bit rate transmission in general. The service interruption may be inevitable when replacing the lower bit rate transmission systems with the higher ones. Another example of the upgrade method by electrical multiplexing is shown in Figure 7.2. This method is based on SCM. The advantage of this method is the minimal use of wavelengths (one or two). The increase in services requires an increase in subcarriers and widens the transmission bandwidth. At the present status of technology, this method imposes system

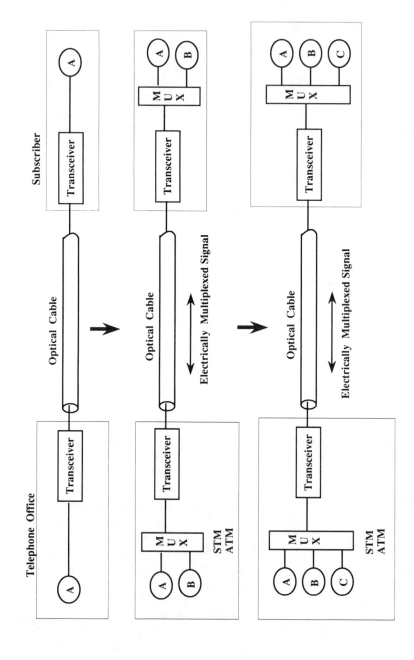

Figure 7.1 Service upgrade by multiplexing.

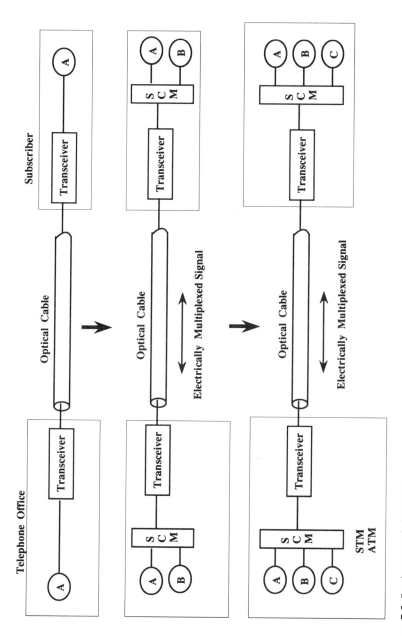

Figure 7.2 Service upgrade by SCM.

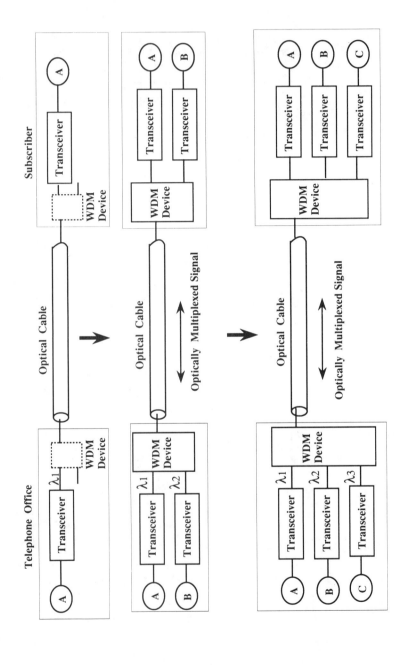

Figure 7.3 Service upgrade by WDM.

interrupt for upgrading. The third example is the service upgrade by WDM, as shown in Figure 7.3. In this method, wavelengths are dedicated to the service, and the number of wavelengths increases along with the increase in services. The advantage of this method is that there is no increase of transmission bit rate. Another advantage is that there is no service interruption for service upgrade when WDM devices are installed initially (as shown in dotted boxes in Figure 7.3).

7.1.2 WDM Devices

The principles of WDM devices are discussed in Chapter 2. Here, these devices are treated from the viewpoint of subscriber loop system design. WDM devices are classified into three types: bulk, fiber, and waveguide types.

The bulk-type device is constructed by the individual components, which are the dispersive component and the coupling component. An example of the dispersive component is a grating on a silicon or on a glass substrate. An example of the coupling component is the rod microlens, and this component realizes the high coupling efficiency between the dispersive component and a single-mode fiber. Even with the coupling component, the coupling loss of bulk-type devices is relatively high, and the coupling mechanism requires precise adjustment in general. The need for precise adjustment is not suitable for mass production. Since individual components are used, the size is relatively large.

The fiber-type device is made by single-mode fibers, which cause the device to have the lowest coupling loss with a single-mode fiber among the three types. This type is good for mass production because of its simple configuration. The cross section of the fused fiber coupler is small but the length is long.

The waveguide-type device is based on the WDM functions in an optical waveguide, which is made from silica or glass materials. The WDM functions are realized by forming a directional coupler or Mach-Zehnder interferometer, or by other means. The waveguide and the formation of WDM functions in a waveguide are made by masking, deposition, and etching processes (similar to the IC production process). Therefore, the waveguide-type device is considered to be suitable for mass production. The waveguide-type device is considered to be the smallest among the devices. When compared to a fused fiber coupler, the coupling loss with a single-mode fiber is not so low (but not so large), because of the spot size difference (mode field difference) between a single-mode fiber and a waveguide.

The characteristics of three types of WDM devices are summarized in Table 7.1. The future device cost will depend on the mass productivity, and this may be the most important factor for subscriber loop systems.

7.1.3 Design Consideration of WDM Systems for Upgrading

Several items must be considered when designing WDM systems, and these are listed in Table 7.2. Wavelengths in WDM systems must be determined by considering the

Table 7.1
Comparison of WDM Devices

Type	Examples	Coupling Loss (SM Fiber)	Size	Mass Production
Bulk	Prism Grating Interferometric optical filter Holography	Passable	Poor to passable	Poor to passable
Fiber	Fused fiber coupler	Excellent	Passable to good	Good
Waveguide	Directional coupler Mach-Zehnder interferometer	Good	Excellent	Excellent

Table 7.2
Design Considerations of WDM Systems

Items	Factors
Wavelength	Fiber loss, device loss, wavelength allocation
WDM device	Cost, performance
Crosstalk	Quality degradation, security
Service	Possibility of interrupt
Bit rate	Loss budget

fiber loss, device loss, available laser diodes, and photodiodes. Optical fiber has a spectral loss as shown in Figure 2.2, and dispersion characteristics as shown in Figure 2.3. These factors limit the possible transmission distance and bit rate. When we use a single-mode fiber, the selection of wavelength influences the normalized frequency parameter V (as indicated in (2.3)). When $V > V_c$ (V_c is the cutoff frequency), multimode transmission is possible even in a single-mode fiber. In system design, the cutoff wavelength λ_c is often used instead of V_c. The relation of (2.3) determines λ_c from V_c. When using a single-mode fiber with $\lambda_c = 1.2$ μm, multimode transmission takes place in the case of the transmission with a wavelength of $\lambda = 0.8$ μm. In multimode transmission, there is an additional noise, which is called *modal noise* [1–3]. When we use a coherent light source such as a laser diode, there are speckle patterns made by the coherent interference among modes in a fiber. In the case that there is spatial filtering in transmission lines, such as optical connectors with lateral displacements, the transmission performance depends on the speckle pattern characteristics. The modal noise originates the random variation of received optical intensity with time as the speckles are varied by several external effects, such as mechanical vibrations. This modal noise will have a serious influence in analog systems, where a high C/N is required, as discussed in Chapter 3. Possible lasing wavelengths by a laser diode and possible detectable wavelengths by a photodiode (as indicated in Figure 2.16) must be taken into consideration in the design of wavelengths in WDM systems.

When selecting WDM devices, cost and performance must be considered. Mass productivity will determine the cost at the massive usage stage. Performance includes insertion loss, reflection loss, and crosstalk.

Crosstalk among channels degrades the transmission quality, which increases the error rate in digital systems and generates the waveform distortion in analog systems. Crosstalk also produces security or privacy problems, in which one person's talk or data can be received by another person. Therefore, crosstalk must be suppressed so that is under the detectable optical level at the bit rate. Data encryption is another solution of this problem.

Services transmitted by WDM systems may influence the system design. When the interruption of the initial service is possible at the service upgrade, the upgrade can be easily performed by the method indicated in Figure 7.3. If the initial service cannot be interrupted, WDM devices must be installed at the initial stage, as indicated by the dotted boxes in Figure 7.3. This increases the cost of the initial system. Besides the laser oscillation and photodetection, a possible optical amplification wavelength must be taken into consideration for some applications. For example, it is better to use the 1.5-μm wavelength region for video distribution services, because the video distribution may require an optical amplifier and the 1.5-μm band is currently available for EDFA amplification.

The signal bit rate is also an important factor for WDM system design. WDM systems use the same optical transmission line, which is divided for each channel through WDM. Therefore, the optical transmission loss is similar to each channel except for the spectral loss dependence of optical fibers and optical devices. High bit rate transmission systems require higher optical received power. Roughly speaking, the following equation for the loss budget holds:

$$\Delta L = 10 \log(B_1/B_2) \quad \text{(dB)} \tag{7.1}$$

where ΔL, B_1, and B_2 are the loss difference in dB and the bit rates for WDM system #1 and #2, respectively. For example, B_1 = 500 Mb/s and B_2 = 50 Mb/s, and then ΔL = 10 dB. If we assign the #1 system to the 1.5-μm region and the #2 system to the 1.3-μm region, the transmission loss for the 1.5-μm region may be smaller. To evaluate the transmission loss difference, we assume the optical transmission losses to be 1 and 0.5 dB/km for 1.3- and 1.5-μm wavelengths, respectively. For 5-km transmission, the transmission loss difference is 2.5 dB, which is not sufficient for $\Delta L = 10$ dB.

One solution to this problem is to design the loss budget for the optical transmission line by considering the highest bit rate when all the services are previously known. Although this is one possible solution, optical subscriber systems depend on each subscriber's service demand, which is generally not known at the system design stage. Therefore, other solutions must be sought for general use. Several countermeasures for this problem are shown in Figures 7.4 and 7.5. These are based on two simple principles: the uses of equivalent high power source and the equivalent low-loss transmission line for the higher bit rate. The use of a high-power laser or optical amplifier corresponds to the equivalent high-power source. The available power of this method is limited by several nonlinear effects, as discussed in Chapter 3. The equivalent low-loss transmission line can be realized by special network configurations, as shown Figure 7.5. One method is the use of a special optical coupler, with which the coupling ratio is different for two systems. A nonequivalent coupling ratio, such as 10:1, realizes the low-loss line for higher bit rate systems. Another method is the use of two branch points in a network; only one branching is used for higher bit rate services [4]. The branching loss for the equivalent power branch is

$$L_B = -10 \log_{10}(1/N) \quad (\text{dB}) \tag{7.2}$$

where N is the branching number. When $N = 8$, then L_B is nearly equal to 9 dB. Although the countermeasures shown in Figures 7.4 and 7.5 are not all-powerful, they may be useful for limited practical applications.

Wavelengths near $\lambda = 1.3$ and 1.55 μm are frequently selected for many systems. Systems using nearly 1.55 μm have the advantages of the lowest fiber loss, and the possible use of the EDFA. System using nearly 1.3 μm have the advantage of minimum dispersion in an ordinary single-mode fiber.

7.2 WDM-BASED MULTICHANNEL SYSTEMS

The systems discussed in the previous section do not use WDM at the initial stage, and WDM is later used as an upgrading method. WDM-based multichannel systems are another application of WDM, and they are intended to use WDM from the initial stage. Several systems have been proposed. One example is shown in Figure 7.6(a), where each wavelength is dedicated to each subscriber [5]. This system requires many wavelengths corresponding to the number of subscribers connected through WDM (e.g., the number is 8 to 16). The advantage of this system is the independence of channels when the level of crosstalk is practically very low. Since this

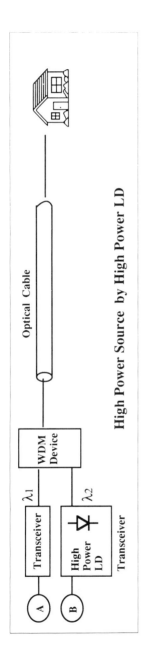 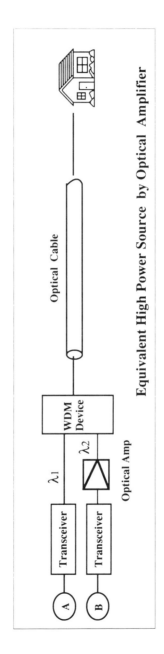

Figure 7.4 Countermeasures for high bit rate, wide-band service upgrade (1).

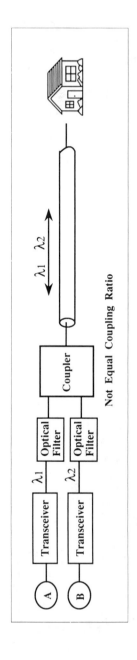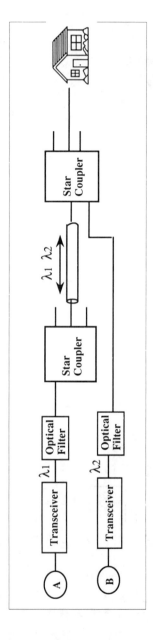

Figure 7.5 Countermeasures for high bit rate, wide-band service upgrade (2).

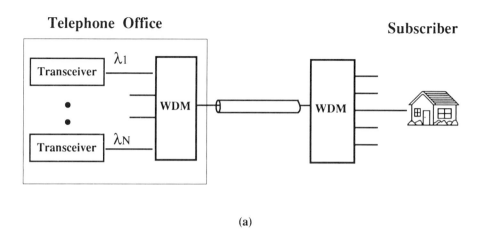

(a)

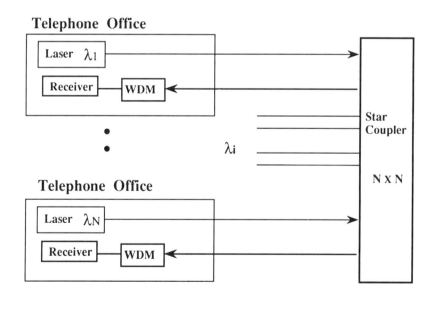

(b)

Figure 7.6 WDM-based systems.

system uses the WDM device, there is no additional loss in principle. Although there is some insertion loss of WDM devices, the transmission loss is very small when compared to a system with a star coupler. The disadvantage of this system is the lack of wavelength transparency and the difficulty of using the new wavelength, which is not considered at a design stage. Another disadvantage is the use of many types of lasers with different wavelengths, which may raise the system cost.

Another example of WDM-based systems is shown in Figure 7.6(b). This system is intended for video distribution applications [6]. The $N \times N$ star coupler is used for mixing the data and is also used as a distribution of the mixed data. This $N \times N$ star coupler should have no wavelength dependence. Because the single power is divided into $1/N$ through the $N \times N$ star coupler with equal division, the loss is also expressed by (7. 2). Each telephone office has a transmitter with a unique wavelength. Other telephone offices can receive this wavelength through a star coupler and a WDM device. When using a fixed WDM device, the predetermined wavelength can be received. With a tunable WDM device, the desired wavelength (not a predetermined wavelength) can be received. One example of a tunable WDM device is shown in Figure 7.7. A grating is used for a WDM device, and a desired wavelength can be detected by rotating this grating.

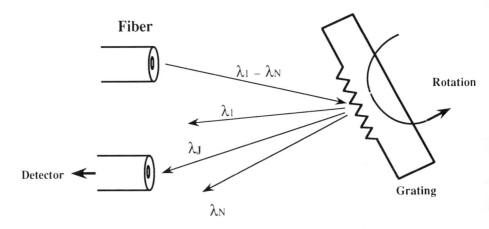

Figure 7.7 Example of tunable WDM devices.

7.3 WDM- AND SCM-BASED MULTICHANNEL SYSTEMS

Wavelength spacing in a WDM system is not so narrow when compared to an OFDM system. For example, wavelength spacing in WDM and the available wavelength range are assumed to be 2 nm and 1.280 to 1.330 μm. Then the possible number of channels is limited to 25. One method to overcome this limitation is the use of

OFDM instead of WDM, which will be discussed in Chapter 8. Another method is the combination of WDM and SCM. When compared to the OFDM approach, this approach is easier to realize at present because OFDM requires sophisticated optical devices and precise wavelength stabilization. In OFDM, the spacings of the wavelengths (optical carrier) are much narrower and precise wavelength stabilization is required. Although intensive investigations have been made, the precise stabilization is not very easy at present. In WDM, no wavelength stabilization is required in general. The stabilization of electrical carriers in SCM systems is well established by using components from a commercial source.

Examples are shown in Figures 7.8 and 7.9, which are proposed and discussed in [7,8]. The system shown in Figure 7.8 at (i, j) station uses the fixed wavelength λ_i and the fixed subcarrier f_j. An optical filter is installed at the network side, and this filter selects the predetermined wavelength. The rf filter in a receiver selects the desired subcarrier. The system shown in Figure 7.9 uses the variable λ_i and f_j, making the network more flexible. The optical filter is installed at the individual stations. By varying the combinations of λ_i and f_j, this network supports several services, such as point-to-point services and broadcast services.

When these two systems have a single channel (a single subcarrier) per laser, the large optical modulation index per channel is possible, and this results in a larger loss budget for the system. When the combination of λ_i and f_j such as (λ_i, f_j) is uniquely dedicated to the (i, j) station, the possible number of channels is increased to NM for N wavelengths and M subcarriers, while the number is N for pure WDM systems. These features are suitable for LANs or *metropolitan area networks* (MAN).

There are limitations to these systems due to optical beat interference, as indicated in [8,9]. Optical beat interference is the result of multiple lasers, as shown in Figure 7.10. In these systems, the wavelength λ_i is shared by M stations, and the number of lasers for λ_i is M. Practically, each laser has the same or a slightly different wavelength, and these wavelengths for λ_i can generally pass through the optical filter. The lights from multiple lasers interfere, and this optical beat results in noise. The current output from a photodiode $i(t)$ is derived from (2.35) and

$$i(t) = R\, P_{\text{total}} \qquad (7.3)$$

where R and P_{total} are the responsivity and the total optical power, respectively. P_{total} is proportional to the square of the total electric field $E(t)$:

$$P_{\text{total}} \propto |E(t)|^2 \qquad (7.4)$$

$$E(t) = \sum_{j=1}^{M} e_j(t) \qquad (7.5)$$

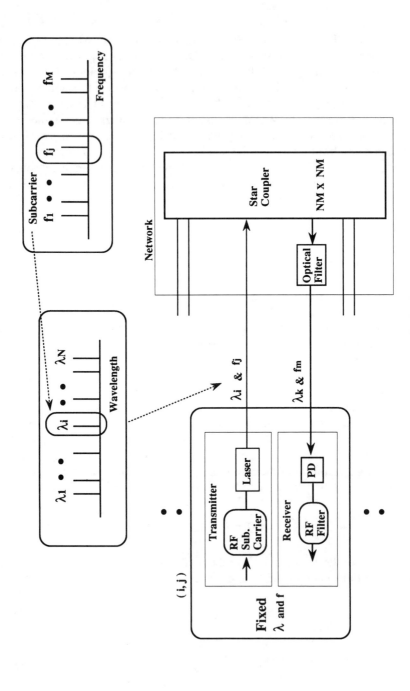

Figure 7.8 WDM- and SCM-based systems (1).

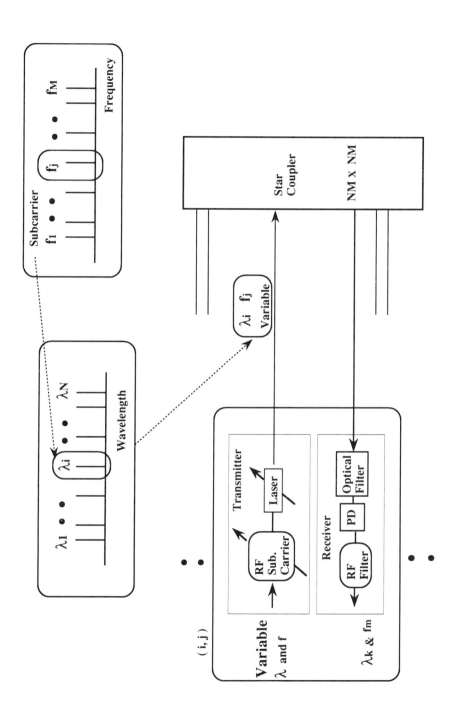

Figure 7.9 WDM- and SCM-based systems (2).

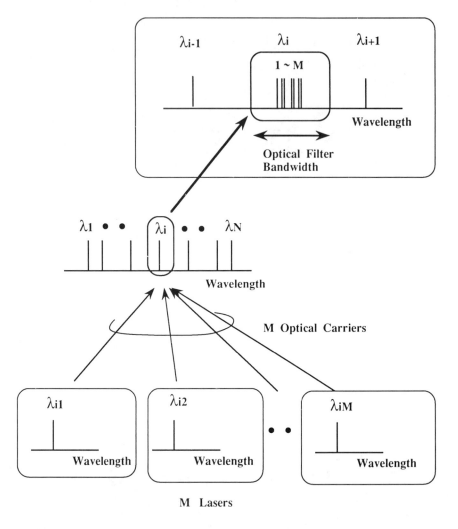

Figure 7.10 Optical beat interference.

where $e_j(t)$ is the individual electric field for each laser with nearly λ_i wavelength. $|E(t)|^2$ is

$$|E(t)|^2 = \sum_j |e_j(t)|^2 + \sum_{j \neq s} [e_j(t)\, e_s^*(t)] \qquad (7.6)$$

where $e_s^*(t)$ is the complex conjugate of $e_s(t)$. Here, the term $\sum_{j \neq s}[e_j(t)e_s^*(t)]$ is composed of $[e_j(t)e_s^*(t) + e_s(t)e_j^*(t)]$. It equals the real part of $[2e_j(t)e_s^*(t)]$. As a result of these calculations, $i(t)$ is [9]

$$i(t) = R\left[\sum_j P_j(t) + \sum_{j \neq s} \sqrt{P_j(t)\,P_s(t)}\cos(2\pi\,\delta f_{js} + \Delta\phi_{js})\right] \quad (7.7)$$

$$\delta f_{js} = c\left(\frac{\lambda_{ij} - \lambda_{is}}{\lambda_{ij}\,\lambda_{is}}\right) \quad (7.8)$$

$$\Delta\phi_{js} = \phi_{ij} - \phi_{is} \quad (7.9)$$

where P_j, c, and ϕ_{ij} are the optical power for the wavelength λ_{ij}, the light velocity, and the phase for λ_{ij}, respectively. The second term of (7.7) expresses the interference, and this is the additional source of noise for these systems.

REFERENCES

[1] Epworth, R. E., "The Phenomenon of Modal Noise in Analogue and Digital Optical Fibre Systems," *Proc. 4th European Conference of Optical Communication (ECOC'78)*, 1978, p. 492.
[2] Hill, K. O., Y. Tremblay, and B. S. Kawasaki, "Modal Noise in Multimode Fiber Links: Theory and Experiment," *Opt. Lett.*, Vol. 5, 1980, p. 270.
[3] Pertermann, K., "Nonlinear Distortions and Noise in Optical Communication Systems Due to Fiber Connectors," *IEEE J. Quantum Electron.*, Vol. QE-16, 1980, p. 761.
[4] Reeve, M. H., S. Hornung, L. Bickers, P. Jenkins, and S. Mallinson, "Design of Passive Optical Networks," *Br Telecom. Technol. J.*, Vol. 7, 1989, p. 89.
[5] Wagner, S. S., H. Kobrinski, T. J. Robe, H. L. Lemberg, and L. S. Smoot, "Experimental Demonstration of a Passive Optical Subscriber-Loop Architecture," *Electron. Lett.*, Vol. 24, 1988, p. 344.
[6] Kobrinski, H., R. M. Bulley, M. S. Goodman, M. P. Vecchi, C. A. Brackett, L. Curtis, and J. L. Gimlett, "Demonstration of High Capacity in the Lamdanet Architecture: A Multiwavelength Optical Network," *Electron. Lett.*, Vol. 23, 1987, p. 824.
[7] Liew, S. C., and K. W. Cheung, "A Broad-Band Optical Network Based on Hierarchical Multiplexing of Wavelengths and rf Subscribers," *IEEE J. of Lightwave Technol.*, Vol. 7, 1989, p. 1825.
[8] Shankaranarayanan, N. K., S. D. Elby, and K. Y. Lau, "WDMA/Subscriber—FDMA Lightwave Networks: Limitations Due to Optical Beat Interface," *IEEE J. of Lightwave Technol.*, Vol. 9, 1991, p. 931.
[9] Desem, C., "Optical Interference in Lightwave Subcarrier Multiplexing Systems Employing Multiple Optical Carriers," *Electron. Lett.*, Vol. 24, 1988, p. 50.

Chapter 8
Optical Transmission Systems Using Coherent Technology

There are several proposed coherent-technology-based systems for subscriber loop applications, and they are divided into two categories. One is the OFDM system and the other is the optical heterodyne system. OFDM systems are the straightforward extension of WDM systems, the difference being the much narrower wavelength spacing in OFDM systems. Although the superior receiver sensitivity is also beneficial to subscriber systems, proposed optical heterodyne systems for the subscriber loop are mainly based on the superior optical frequency selectivity (wavelength selectivity) of optical heterodyne technology. The key devices for these systems are tunable laser diodes and optical multi/demultiplexers. In this chapter, proposed systems and these key devices are explained.

8.1 TUNABLE LASER DIODE

Tunable laser diodes are those laser diodes whose lasing wavelength can be tuned to a desired wavelength. The tuning can be realized with several methods, such as varying injection current, varying environment temperature, or using the wavelength selection functions (e.g., by rotating a grating in an external cavity laser). Tuning by temperature can be applied for all lasers because this method is based on the temperature shift of lasing wavelength (the shift is about 0.05 to 0.5 nm/deg). Although this method is useful at the laboratory level for some experiments, it is limited for practical application because of its slow tuning speed. Therefore, a discussion of temperature tuning is not included here. Tunable laser diodes can be used as main light sources with several different wavelengths in OFDM systems or as a local oscillator in heterodyne-based subscriber systems.

8.1.1 Multielectrode Tunable Laser Diode

The multielectrode laser structure is commonly used for a tunable laser diode. Typical structures are shown in Figure 8.1 for both the DFB and DBR lasers. The injection current varies the lasing wavelength for both types. The basic principle is the refractive index change in a cavity by the injection current, known as the *plasma effect*. This is expressed as [1]

$$\Delta n = C_N \Delta N \tag{8.1}$$

where Δn and ΔN are the refractive index change and the change in the injected carrier density, respectively. C_N is the constant ($C_N < 0$) for the fixed wavelength and material. The values of C_N are different for the different wavelengths and materials. The effective refractive index change Δn_{eff} is expressed by using the mode confinement factor Γ and (8.1):

$$\Delta n_{\text{eff}} = \Gamma C_N \Delta N \tag{8.2}$$

Since ΔN corresponds to the injection current, the refractive index decreases by increasing the injection current ($C_N < 0$ in (8.2)). It is estimated that the relative index change $\Delta n_{\text{eff}}/n_{\text{eff}}$ is about 1%, and the resulting possible tuning range is around 10 nm. A tunable laser diode was proposed based on this mechanism [2] and the first tunable DBR-LD was fabricated in 1983 [3]. Many improvements have been made, and the important improvements are the carrier injection into a Bragg reflection region [4] for a wide tuning range and the introduction of a phase control region for continuous tuning [5]. The three-electrode DBR laser shown in Figure 8.1 has the improved structure, where active, phase control, and Bragg reflection regions are built into a chip. The maximum reflection in a DBR laser takes place at the following Bragg wavelength λ_B:

$$\lambda_B = 2 n_{\text{eff}} \Lambda \tag{8.3}$$

where Λ is the first-order Bragg grating period. The Bragg wavelength λ_B varies according to (8.3) because of the injected carriers. The carrier density change causes changes in both the refractive index and the gain coefficient. The refractive index and gain coefficient are the real and imaginary parts of the complex refractive index, and they are closely linked by the Kramers-Kronig relation. Therefore, not only the refractive index but also the gain are changed. The ratio of these changes is expressed by the linewidth enhancement factor α, which is discussed in Chapter 2. With a large value of α, large refractive index change is obtained for the same gain change. In the DBR lasers, the band gap in the grating region is large at the lasing wavelength, and this results in a large α value. The band gap in the active region (the

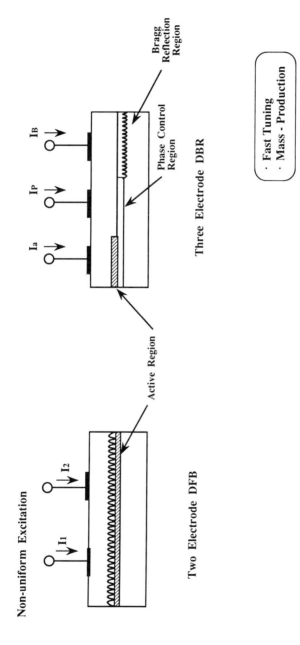

Figure 8.1 Examples of multielectrode tunable laser diode.

large gain region) is small and α is small. Schematically speaking, only the refractive index is changed in the Bragg reflection region by injected carriers while the gain is changed in the active region. Therefore, the lasing wavelength is mainly determined by the lasing condition through (8.3). This inhomogeneity of α is considered to be the substantial tunability of the DBR laser [6].

The first tunable two-electrode DFB lasers were demonstrated in 1986 [7,8]. In DFB lasers, the tuning is realized by the nonuniform injection currents (nonuniform excitation). As indicated in Figure 8.1, the two sections of two-electrode DFB lasers are structurally identical, and the value of α essentially takes the same values in both of the sections. The mechanism of tunability in DFB lasers can be explained by the different effective lengths of the sections and the nonuniform gain [9]. The effective length L_{eff} of a grating is determined by the detuning from the Bragg condition and the gain. The gain is modulated by the grating, and this results in the nonuniform gain. The wavelength shift $\Delta\lambda$ for two electrode DFB lasers is expressed as [9]:

$$\Delta\lambda = \Delta\lambda_{B1}\left(\frac{L_{\text{eff1}}}{L_{\text{eff1}} + L_{\text{eff2}}}\right) + \Delta\lambda_{B2}\left(\frac{L_{\text{eff2}}}{L_{\text{eff1}} + L_{\text{eff2}}}\right) \qquad (8.4)$$

where $\Delta\lambda_{Bi}$ ($i = 1, 2$) is the Bragg wavelength shift for each section, and this is

$$\Delta\lambda_B = \frac{\Delta n_{\text{eff}}}{n_{\text{eff}}}\lambda \qquad (8.5)$$

When the carrier density increases in one section from the current injection, the carrier density in the other section decreases, and then the total gain is balanced with the cavity loss. Therefore, $\Delta\lambda_{B1}$ and $\Delta\lambda_{B2}$ in (8.4) have opposite signs. The effective lengths L_{eff} in (8.4) are different for each section in two-electrode DFB lasers; then the cancellation does not occur according to (8.4), and the wavelength shift (tuning) is realized. Many improvements have been made; these are the introduction of the phase-shift control region [10] and the MQW structure. A three-electrode DFB laser shows a flat FM response from several tens of hertz to several hundred megahertz of modulation frequency [11]. Another tuning laser, the tunable twin-guide laser, has been also proposed [12].

The injected carriers cause not only the refractive index change but also the absorption loss increase. In general, this loss increase makes the lasing threshold increase and the differential quantum efficiency decrease. Fast tuning can be expected in these multielectrode laser diodes because of the current tuning. The obtained tuning speed was around 1 ns for DBR lasers. As for the high-speed tuning, multielectrode DFB lasers have the advantage over multielectrode DBR lasers because of the shorter carrier lifetime due to the stimulated emission. Since these laser diodes are monolithic, they are suitable for mass production.

8.1.2 External Cavity Tunable Laser Diode

There have been several external cavity tunable laser diodes proposed so far. The general structure uses an ordinary laser diode chip and an external cavity, which has the wavelength selectivity, as indicated at the top of Figure 8.2. The laser diode chip is usually AR coated on the external cavity side. It forms the laser cavity between the laser diode chip facet and the wavelength-selective device. Since it is a compound device, it is not very suitable for mass production. On the other hand, it has two major advantages over the present monolithic tunable lasers, such as a large tuning range and a narrow linewidth. The gain range of laser diodes is very wide (e.g., 400 nm), and the tuning range can in principle be equal to this wide gain range for the external cavity tunable lasers. The narrow linewidth is obtained from long cavity length due to an external cavity. It is difficult to obtain continuous tuning because the wavelength-selective device selects one mode among the closely spaced modes, and mode hopping generally occurs in this selection process.

The proposed external cavity tunable lasers can be categorized into three groups, according to the working principles of wavelength selection. They are grating, optical filtering, and compound cavity. In a laser with a grating, the tuning is done by mechanically rotating the grating, which is used for a wavelength-selective device. The very wide tuning range of 242 nm from using the MQW structure [13] and the very narrow linewidth of 10 kHz [14] were obtained from this type of laser diode.

For a laser using acousto-optical filtering, tuning is done by the acoustic frequency f_a driven by an electric frequency [15]. For a given acoustic frequency f_a, a small range of the optical frequencies which satisfy the momentum conservation law [16] can be diffracted, and this acts as an optical filter. Since the diffracted optical beam is frequency-shifted by f_a, an acousto-optic modulator is inserted to correct this frequency shift of the beam. A tuning range of about 70 nm and a tuning speed of 3 μs were reported using a 1.3-μm InGaAsP laser diode [15].

In the case of a laser using electro-optical filtering, tuning is done by the voltage imposed on the TE-TM converter [17]. The optical band-pass filter is constructed with the TE-TM converter and the built-in polarizer. A maximum tuning range of 7 nm was obtained [17].

In a compound cavity laser, the lasing occurs at the wavelength where the resonance wavelengths for two cavities coincide. A small change in one mode wavelength results in a large lasing wavelength change with mode hopping because the compound cavity modes must coincide for lasing. A tuning range of about 30 nm was obtained [18].

8.1.3 Tuning Characteristics of Tunable Laser Diode

Several characteristics of a tunable laser must be considered for system design, such as tuning range, tuning speed, linewidth, and whether or not there is continuous

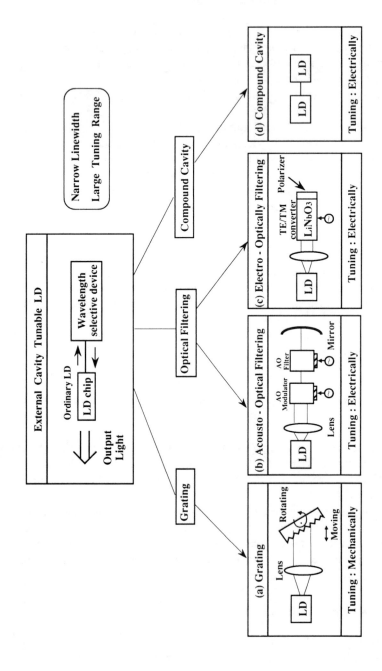

Figure 8.2 Examples of external cavity tunable laser diode.

tuning. The importance of these characteristics is different for the different applications. When a tunable laser is used as a local oscillator in a coherent transmission system, fast tuning is desirable. On the other hand, fast tuning is not always required when it is used as a main light source with various wavelengths. When it is used as a main light source in an OFDM system, the linewidth may be the important characteristic.

Either continuous tuning or quasi-continuous tuning is realized when the tuning currents of the multielectrode tunable laser diodes are varied, as shown in Figure 8.3. Here, continuous tuning is defined as tuning that maintains the same longitudinal mode. Quasi-continuous tuning is defined as continuous tuning with a different longitudinal mode [19]. When the current increases for both tunable DBR and DFB lasers, the wavelength is shifted as indicated in Figure 8.3. At present, about 4-nm continuous tuning and about 8-nm quasi-continuous tuning have been reported for these lasers. Continuous tuning is easy to handle in general.

The linewidth of a tunable laser diode is important for coherent transmission systems (see Figure 8.4). As discussed in Chapter 3, receiver sensitivity is degraded for a broad linewidth. For example, a linewidth $\Delta \nu$ must be under $0.1 R_b$ for a 1-dB loss penalty in a heterodyne FSK system, where R_b is the bit rate. In OFDM systems, crosstalk due to an overlap in optical frequency (wavelength) domain takes place for lasers with a broad linewidth.

8.2 OPTICAL MULTI/DEMULTIPLEXER FOR OFDM

The multi/demultiplexer can be used in various OFDM systems, as shown in Figure 8.5. In Figure 8.5(a,b), an optical demultiplexer (DEMUX) is located at each subscriber and the signals from a telephone office are power-divided by a star coupler, which is not located at each subscriber. In the case of Figures 8.5(c,d), the OFDM signals are demultiplexed by a demultiplexer and only the dedicated wavelength is sent to each subscriber. From the security or privacy point of view, it is better to adopt the system shown in Figure 8.5(c or d). It is possible to use a star coupler instead of an optical multiplexer (MUX) for a source light combiner at the cost of causing dividing loss.

OFDM systems require finer multi/demultiplexers because of the narrower spacing in the optical frequency domain (wavelength) compared to WDM systems. Three types are discussed below, such as a Mach-Zehnder interferometer type, a Fabry-Perot interferometer type, and laser diode filter type. The former two types use the principle of optical interference, which realizes finer multi/demultiplexing. The third type is based on the tuning mechanism of a laser diode.

An interferometer-type device is shown in Figure 8.6, and the place where the interference occurs is indicated in the same figure. As the result of the interference of light, wavelength selectivity is realized and this selectivity appears periodically in the optical frequency (wavelength) domain, as shown in Figure 8.6.

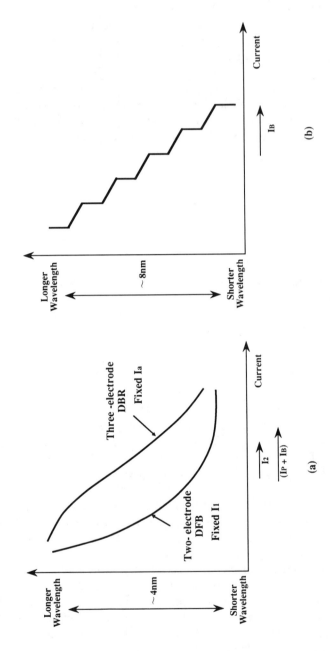

Figure 8.3 Characteristics of tuning: (a) continuous tuning; (b) quasi-continuous tuning.

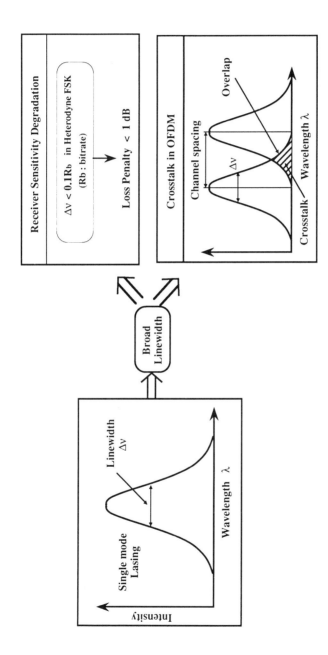

Figure 8.4 Linewidth of tunable laser diode.

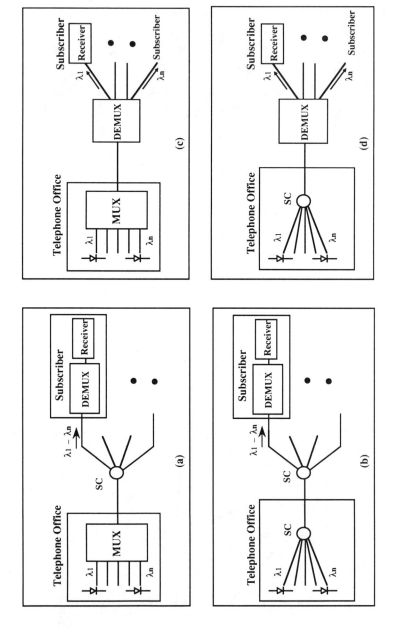

Figure 8.5 Multi/Demultiplexers in OFDM systems: (a,b) where signals from telephone office are power-divided by a star coupler; (c,d) where only the dedicated wavelength is sent.

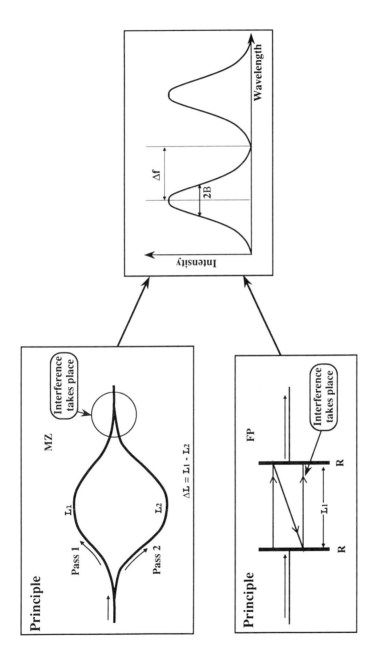

Figure 8.6 Interferometer-type device for OFDM.

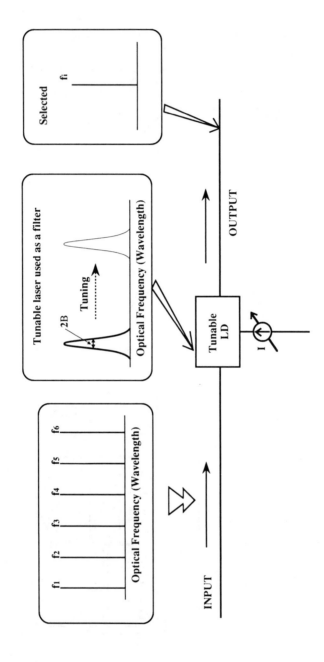

Figure 8.7 Laser diode filter.

The use of a laser diode filter is shown in Figure 8.7 and the input optical frequencies are selected by a laser diode filter. The variable selection of input frequencies is possible when using a tunable laser diode for a laser diode filter.

8.2.1 Mach-Zehnder Interferometer Type

The typical Mach-Zehnder interferometer-type device has been made by forming the interferometer in a waveguide [20]. Although this device can be used as both a multiplexer and a demultiplexer, here we call it a *Mach-Zehnder guided wave demultiplexer* (MZ-GWD) for the sake of simplicity.

Basic Properties

The basic properties of the MZ-GWD have been investigated [21]. Frequency space Δf between the maximum and minimum transmission power of MZ-GWD was derived as

$$\Delta f = v_c/(2\, n_e\, \Delta L) \tag{8.6}$$

where n_e is the effective refractive index of a waveguide, v_c is the light velocity in a vacuum, and ΔL is the length difference in the MZ-interferometer. Extinction ratios, E_1 and E_2, of two outputs for a 1×2 demultiplexer were derived as

$$E_1 = -10 \log[(\sqrt{R_1 T_2} - \sqrt{R_2 T_1})^2/(\sqrt{R_1 T_2} + \sqrt{R_2 T_1})^2] \tag{8.7}$$

$$E_2 = -10 \log[(\sqrt{R_1 R_2} - \sqrt{T_1 T_2})^2/(\sqrt{R_1 R_2} + \sqrt{T_1 T_2})^2] \tag{8.8}$$

where $R_1 = 1 - T_1$ and $R_2 = 1 - T_2$. Two directional couplers are used in a 1×2 demultiplexer. T_1 and T_2 are power transfer ratios of directional couplers. Using these equations, frequency space Δf and extinction ratios E_1 and E_2, can be calculated.

Wavelength Dependence of Δf and Extinction Ratio

The wavelength dependence of the frequency space and extinction ratio is important for some applications, such as the system described in Section 8.3. The wavelength dependence of these parameters cannot be calculated without knowing the wavelength dependence of n_e, T_1, and T_2. For obtaining wavelength dependence, directional couplers in a waveguide are modeled in Figure 8.8. This figure is a cross section (x-, y-planes) with two rectangular cores with refractive index n_1 and cladding with n_2. Each core has the dimensions of $a \times a$. The separation of the two cores is

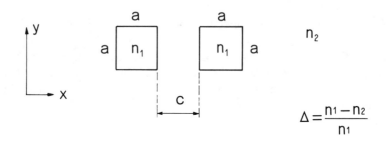

Figure 8.8 Model of a directional coupler in a guided-wave demultiplexer for calculation. Two rectangular cores with refractive index n_1 and cladding with n_2 (© 1991 IEEE).

c. By using the propagation constant for the z-direction k_z and coupling constant K, n_e and T_1 are expressed as follows:

$$n_e = (\lambda/2\pi)k_z \tag{8.9}$$

$$T_1 = \cos^2(KZ) \tag{8.10}$$

where λ and Z are the wavelength in a vacuum and the coupling length, respectively. In these equations, k_z and K for the fundamental mode E_{11} with y polarization can be calculated by the approximate theory in [22] as

$$k_z = (k_1^2 - k_x^2 - k_y^2)^{1/2} \tag{8.11}$$

and

$$K = 2Ak_x^2[1 - (k_x A/\pi)^2]^{1/2} \exp\{-\pi c[1 - (k_x A/\pi)^2]^{1/2}/A\}/(\pi a k_z) \tag{8.12}$$

where A, k_1, k_z, and k_y are expressed as

$$A = \lambda/(n_1^2 - n_2^2)^{1/2} \tag{8.13}$$

$$k_1 = (2\pi/\lambda)n_1 \tag{8.14}$$

$$k_x = \pi/(a + 2A/\pi) \tag{8.15}$$

$$k_y = \pi/(a + 2An_2^2/\pi n_1^2) \tag{8.16}$$

The wavelength dependence of n_1 and n_2 for a silica waveguide is assumed to be that of optical-fiber glass. The wavelength dependence of refractive index n is generally expressed as the following Sellmeier equation:

$$n^2 - 1 = \frac{a_1\lambda^2}{\lambda^2 - b_1} + \frac{a_2\lambda^2}{\lambda^2 - b_2} + \frac{a_3\lambda^2}{\lambda^2 - b_3} \tag{8.17}$$

where a_i and b_i ($i = 1 \sim 3$) are constants. The constants of the Sellmeier equation for fiber glass were obtained [23], and these were used for calculations. Similar expressions for fundamental mode E_{11} with x polarization can be derived (but are not shown here).

Using (8.6) to (8.17) and refractive-index wavelength dependence, frequency space Δf and the extinction ratio were calculated for a 1.5- to 1.6-μm wavelength [24] and are shown in Figure 8.9(a,b). The parameters are taken as $c = 2$ μm, $\Delta = 0.5\%$, and $c = 3$ μm, $\Delta = 0.8\%$, where Δ is defined as $(n_1 - n_2)/n_1$. The worst value between two extinction ratios is shown in Figure 8.9(b). It is assumed in the calculations that $\Delta f = 5$ GHz and the extinction ratio $= \infty$ at a 1.55-μm wavelength. The calculated results show that Δf is almost constant for a 1.5- to 1.6-μm wavelength, while the extinction ratio depends on wavelength.

The measured values of Δf and the extinction ratio are also shown in Figure 8.9(a,b) as circles. A tunable DBR-LD with two electrodes was used as a wavelength-variable light source. The wavelength of the tunable laser was swept over several gigahertz by varying the injection current into the Bragg region. One example of the measured output is shown in Figure 8.10. The swept frequencies were measured by a Fabry-Perot etalon-based optical spectrum analyzer. Three tunable DBR-LDs were used in the measurements. The periodic property of MZ-GWD was confirmed for 14 nm (about 1750 GHz) experimentally. The measured extinction ratios are almost constant for the wavelength variation and the value is small. This discrepancy between the calculated and measured values is explained as follows: In calculations, extinction ratio $= \infty$ at a 1.55-μm wavelength is assumed. This means that the special relations such as $T_1 = T_2$ must be held for two directional couplers in MZ-GWD. However, currently, these relations are not realized exactly in an actual MZ-GWD. When manufacturing techniques of MZ-GWD improve, higher extinction ratios will be realized and the discrepancy will be decreased.

Tunability

By forming a thin-film heater in one arm of the MZ-interferometer, tunable filtering was realized. By using a two-stage MZ-GWD, a 5-GHz-spaced eight-channel tunable filter with a tuning speed of 0.8 ms was reported [25].

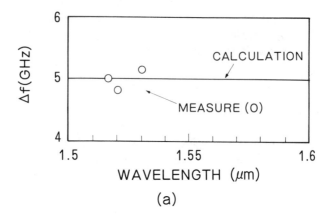

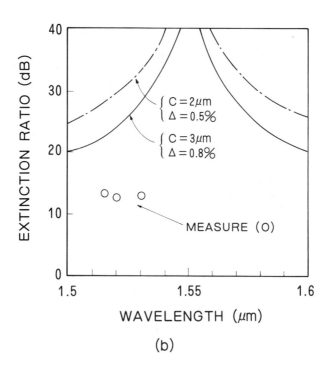

Figure 8.9 Calculated and measured frequency space of (a) Δf and (b) extinction ratio as a function of wavelength. Circles represent measured values. The parameters are as indicated in the figures, where Δ is defined as $(n_1 - n_2)/n_1$. (© 1991 IEEE.)

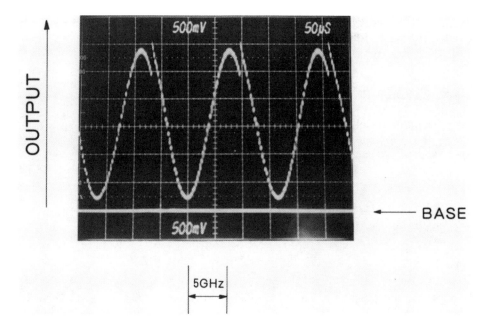

Figure 8.10 Measured output of a guided-wave demultiplexer as a function of an optical frequency (© 1991 IEEE).

8.2.2 Fabry-Perot Interferometer Type

Basic Properties

The basic structure of a Fabry-Perot interferometer is shown in Figure 8.6, and this is the same as that in Figure 5.5 (optical amplifier) except for the gain. Therefore, the derived equation (5.4) can be applied with $G_s = 1$ (no gain) and $R_1 = R_2 = R$, and then

$$G_T = \frac{(1-R)^2}{(1-R)^2 + 4R\sin^2(\beta L)} \tag{8.18}$$

Here, we define Δf as the frequency interval between the maximum and minimum transmission power (the definition of Δf is different from that in (5.8)); then Δf is

$$\Delta f = v_c/(4 n_e L) \tag{8.19}$$

Bandwidth B, defined as half width at half maximum (HWHM), is

$$B = \frac{v_c}{2\pi n_e L} \sin^{-1}\left(\frac{(1-R)}{2\sqrt{R}}\right) \tag{8.20}$$

The ratio of Δf and B is commonly used as a measure of an FP filter and the ratio is called finesse F. Finesse F is

$$F \equiv \frac{2\Delta f}{2B} = \frac{\pi}{2}\left[\sin^{-1}\left(\frac{1-R}{2\sqrt{R}}\right)\right]^{-1} \qquad (8.21)$$

and F is approximately expressed for $R \sim 1$:

$$F \sim \frac{\pi\sqrt{R}}{1-R} \qquad (8.22)$$

For example, $F = 312$ for $R = 99\%$. For obtaining a high-finesse filter, two-stage FP filters were considered for OFDM systems [26].

Tunability

By adjusting the separation length, a tunable FP filter can be realized. There are several mechanisms for adjusting, and one example is the use of a piezoelectric control device.

8.2.3 Laser Diode Filter Type

A laser diode with a single longitudinal mode such as a DBR or DFB laser can be used as an optical filter. The laser biased at a level slightly below the threshold acts as an optical selective amplifier. It only amplifies the input optical signals with the optical frequency, which coincides with the unique frequency determined by the laser. This is similar to an FP optical amplifier, except that a DBR or DFB LD filter amplifies only one frequency (mode). An FP amplifier can amplify many modes, as shown in Figure 5.7. The first LD filter using a DFB laser was demonstrated in 1987, and a selectivity of $\Delta f = 9$ GHz and a tunability of 3.8 GHz/mA were obtained [27]. By using a multielectrode DFB laser, a tuning range of 120 GHz with a 24.5-dB constant gain was achieved [28]. The advantages of LD filtering are optical gain and easy tuning (by current). However, an LD filter used as a demultiplexer can only be applied to the system of Figure 8.5(a,b).

8.2.4 Other Types

Optical filtering devices used for an external cavity tunable laser diode (see Figure 8.2) are also used for an optical filter for OFDM. They are acousto-optical filtering and electro-optical filtering, which have already been explained.

The nonlinear effect of SBS in a single-mode fiber is used for an optical filter [29]. Brillouin amplification takes place only for the input signal, which has a shifted frequency of 11 GHz (Brillouin shift for 1.5-μm light in a silica fiber) from a pump frequency. By changing the frequency of the pump laser, tunable filtering is realized.

8.3 PASSIVE OPTICAL NETWORK USING OFDM

Here, one network configuration and related experiments are shown as an example of an OFDM-based passive optical network.

8.3.1 Network Configuration

Here, we discuss an OFDM-based passive optical network, as shown in Figure 8.11 [30]. When using WDM, widely spaced wavelengths are assigned for both upstream and downstream transmission. Therefore, the emitters (laser diodes) must be slightly different, since they emit different light wavelengths. This could prohibit mass production and result in a high cost. To avoid this, closely spaced optical frequencies are used (OFDM), which can be emitted by one type of tunable laser. A *wavelength-selective star coupler* (WSSC) is used between the *central office* (CO) and the subscribers. For downstream transmission, closely spaced (e.g., 5 GHz) wavelengths λ_{11} to λ_{1N} are used based on the OFDM technique. These wavelengths are separated by the WSSC. In downstream transmission, each wavelength is uniquely assigned to each subscriber for privacy, and each subscriber receives only his or her information. Wavelength λ_{11} to λ_{1N} are emitted by one transmitter (TX) at the CO. The same type of receiver can be used for the subscribers because the receiver sensitivity is nearly equal as a result of the closely spaced wavelengths. For upstream transmission, one wavelength λ_0 is used for all subscribers. Therefore, all subscribers can use the same type of transmitter. Furthermore, wavelength λ_0 ranges over a relatively wide spectrum (e.g., from 1.29 to 1.33 μm). This results in low cost because of easy mass production. Each subscriber's information is transmitted through the WSSC to the CO using time position multiplexing to avoid collisions.

The basic network configuration, as shown in Figure 8.11, has several variations according to the devices used. One possible implementation is shown in Figure 8.12. A GWD and a star coupler are used for the WSSC. A tunable laser is used as the CO transmitter, which emits λ_{11} to λ_{1N} wavelengths. The output of this tunable laser is shown schematically in Figure 8.13(a). Wavelengths λ_{11} to λ_{1N} are periodically emitted for a fixed number of time slots. These are separated by the WSSC, as shown in Figure 8.13(b). For each subscriber, information is regularly transmitted, but in bursts. Privacy for each subscriber is achieved totally when no crosstalk oc-

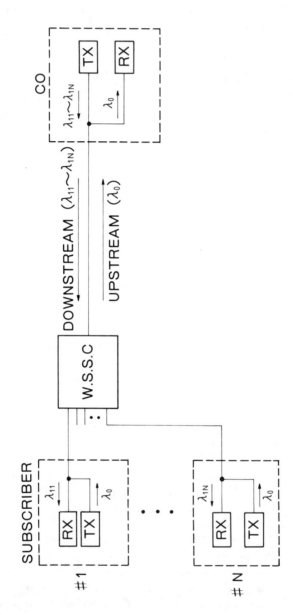

Figure 8.11 Basic configuration of the proposed PASS-NET. WSSC, TX, and RX represent wavelength-selective star coupler, transmitter, and receiver, respectively (© 1990 *J. Opt. Comm.*).

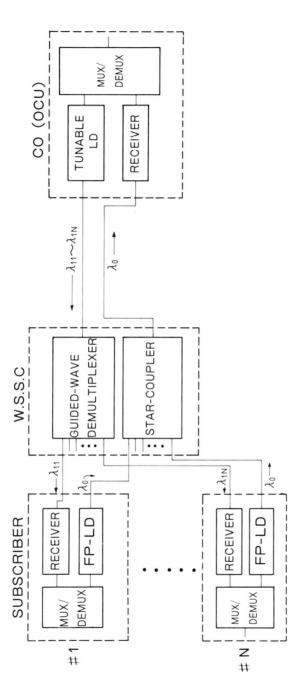

Figure 8.12 One configuration for PASS-NET implementation. The WSSC is composed of a guided-wave demultiplexer and a star coupler (© 1990 *J. Opt. Comm.*).

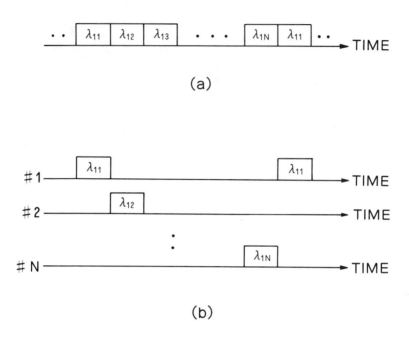

Figure 8.13 Outputs for downstream transmission: (a) outputs for tunable laser; (b) outputs for guided-wave demultiplexer (© 1990 *J. Opt. Comm.*).

curs. For upstream transmission, an ordinary FP-LD is used by each subscriber and the star coupler combines this transmitted light. To avoid collisions, preassigned TDMA is adopted, as shown in Figure 8.14. The output from each subscriber is shown in Figure 8.14(a). Time slots for individual subscribers are preassigned. The receiver at the CO receives light as shown in Figure 8.14(b). To absorb length differences between the star coupler and the subscribers, guard time T_G is set between the individual subscriber's time slots. An IM scheme is used in both upstream and downstream transmissions.

This network is basically designed for narrowband services (such as N-ISDN). After implementation, the network must be smoothly and economically upgraded to broadband services. For distribution services, a possible upgrade plan is shown in Figure 8.15. The services are distributed from the CO with wavelength λ_2. The star coupler has a wavelength-independent property for the wavelength region in use and can transmit both λ_0 and λ_2. To avoid the interruption of narrowband services, which are represented by λ_0 and λ_{11} to λ_{1N}, some devices must be initially installed, as shown in Figure 8.15. WD stands for a *wavelength demultiplexer*, which is used to separate two wavelengths (λ_0 and λ_2). In this figure, subscriber #1 receives both services, while subscriber #N receives only narrowband service. Subscriber #J hopes to receive distribution service in the future. Possible wavelength allocation is shown

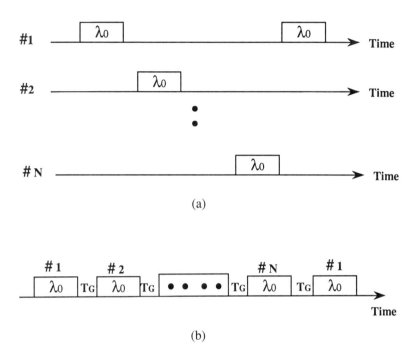

Figure 8.14 Outputs for upstream transmission; (a) and (b) are outputs for each subscriber and star coupler; T_G = guide time (© 1990 *J. Opt. Comm.*).

in Figure 8.16(a). Wavelengths λ_{11} to λ_{1N} are spaced close to each other, while λ_0 and λ_2 are widely separated to allow for ordinary WDM technology.

Another upgrade plan for broadband services is shown in Figure 8.17. This plan is designed to transmit broadband information both ways. Possible wavelength allocation in this case is shown in Figure 8.16(b). Wavelengths λ_{21} to λ_{2N} are closely spaced not only to each other, but also to λ_{11} to λ_{1N}. A GWD using an MZ-interferometer periodically transmits spaced wavelengths. For example, the port for λ_{11} can transmit λ_{21} and cannot transmit λ_{12} to λ_{1N} and λ_{22} to λ_{2N}. Wavelengths λ_{21} to λ_{2N} are reserved for subscribers #1 to #N. To avoid the interruption of narrowband service, some devices must be initially installed as shown in Figure 8.17. In this case, the WD of a subscriber is used to separate two wavelengths, such as λ_{11} and λ_{21}. In this figure, subscriber #1 receives both services, while subscriber #N receives only narrowband service. Subscriber #J hopes to receive broadband service in the future. Transmitters, which emit λ_{21} to λ_{2N} wavelengths, can be set by one type of tunable laser. A fixed wavelength can be set by a fixed current injected into the tunable region in a tunable laser.

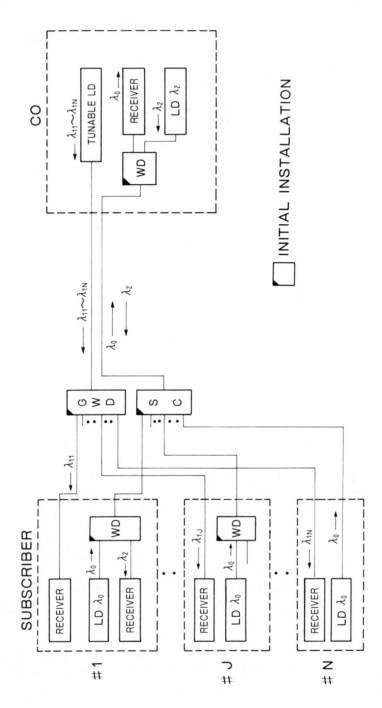

Figure 8.15 One possible configuration to upgrade PASS-NET to distribution services. GWD = guided-wave demultiplexer. SC = star coupler. WD = wavelength demultiplexer (© 1990 *J. Opt. Comm.*).

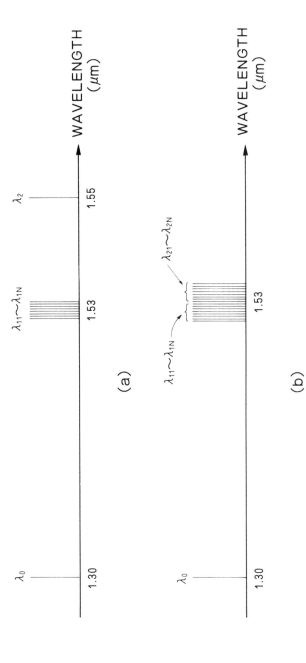

Figure 8.16 Examples of wavelength allocations: (a) corresponds to the configurations shown in Figure 8.15; (b) corresponds to the configurations shown in Figure 8.17 (© 1990 *J. Opt. Comm.*).

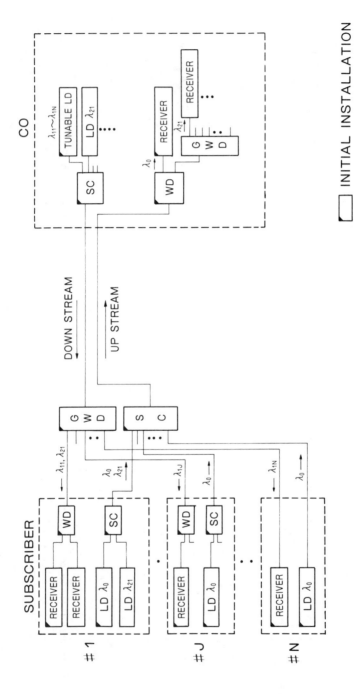

Figure 8.17 One possible configuration to upgrade PASS-NET to broadband services. GWD = guided-wave demultiplexer. SC = star coupler. WD = wavelength demultiplexer (© 1990 *J. Opt. Comm.*).

8.3.2 Experiments

Based on the configuration in Figure 8.17, fundamental experiments for downstream transmission were executed for three waves (two subscribers, $N = 2$) [24]. The experimental setup is shown in Figure 8.18. Two tunable lasers (DBR-LD) with a 1.53-μm wavelength were used as a light source, one for narrowband transmission and another for broadband. Two sets of data, one a pseudo-random pulse ($2^{15} - 1$) at 768 kb/s and the other a dummy (made by using clock pulses), were multiplexed at 32 Mb/s by a multiplexer. The emitted wavelengths were intensity modulated directly at 32 Mb/s. The third wavelength emitted by another tunable laser with an 11-nm separation from the two wavelengths (wavelength was around 1.519 μm) was intensity modulated directly at 100 Mb/s and overlaid. The data at 100 Mb/s were a pseudo-random NRZ bit stream with a $2^{15} - 1$ pattern length. By controlling the injection current into the Bragg region, the third wave could be λ_{21} or λ_{22}. Three waves ($\lambda_{11}, \lambda_{12}, \lambda_{21}$ or $\lambda_{11}, \lambda_{12}, \lambda_{22}$) transmitted over a 10-km single-mode fiber were demultiplexed by a 1×2 GWD. This GWD separated the 5-GHz spaced wavelengths. The GWD used is sensitive to the polarization of the input light. Therefore, a polarization controller (PC) is used. One subscriber received two waves (e.g., λ_{11}, λ_{21}), one for a 32-Mb/s burst transmission and another for a 100-Mb/s continuous transmission. These waves were demultiplexed by a WD. For 100-Mb/s data, the error rate was measured directly. On the other hand, the burst data (narrowband) were converted into continuous data by a demultiplexer and then the error rate was measured for the converted continuous 768-kb/s data.

The first experiment concerns three waves: λ_{11}, λ_{12}, and λ_{22}. The generated wavelengths are shown in Figure 8.19. These photographs were the outputs of a 1×2 GWD. In Figure 8.19(a), only two wavelengths, λ_{11} and λ_{12}, were generated. The upper trace in this photograph corresponds to λ_{12}, while the lower trace to λ_{11}. The data in λ_{11} were short bursts, while the dummy data in λ_{12} were long bursts. The third wave, λ_{22}, was overlaid on the output arm of λ_{12}. The data of this third wave were continuous 100-Mb/s data. The measured outputs for the overlaid signals are shown in Figure 8.19(b). The waves were measured by an optical spectrum analyzer and the results are shown in Figure 8.20. Figure 8.20(a,b) corresponds to two time-averaged outputs from a 1×2 GWD, respectively. The crosstalk of λ_{22} is observed in Figure 8.20(b). Crosstalk is defined here as the ratio of optical transmission peak power to the optical input peak power. The measured error rates as a function of received peak power for 32-Mb/s burst data (λ_{11}) are shown in Figure 8.21. To investigate the degradation caused by the upgrade (overlay), the error rate without λ_{22} was also measured. Circles and crosses stand for data measured with and without λ_{22}. Comparing the two error rate's curves, a 0.7-dB degradation is observed for 10^{-9}. This is due to the interference caused by crosstalk. The suppression of the wavelength fluctuation of a laser and the realization of GWD with a higher extinction ratio is important for lowering the crosstalk level.

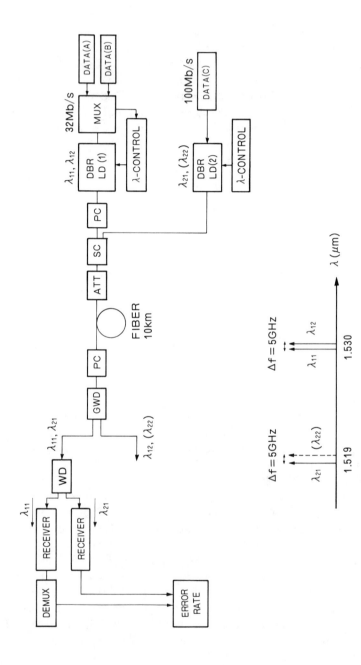

Figure 8.18 Setup for upgrade experiments GWD = guided-wave demultiplexer. SC = star coupler. WD = wavelength demultiplexer (© 1991 IEEE).

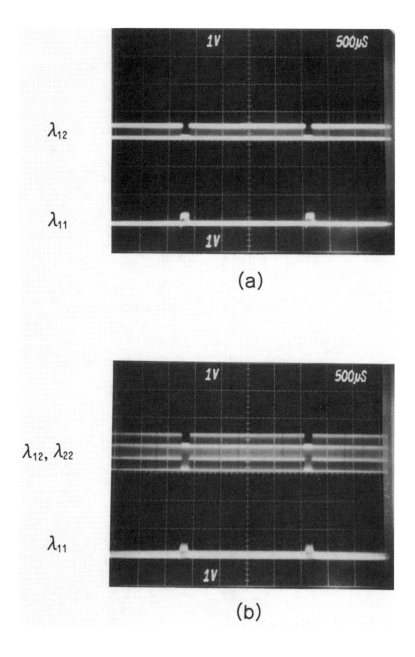

Figure 8.19 Outputs of a guided-wave demultiplexer: (a) generated two waves at 32 Mb/s with a tunable laser; upper trace corresponds to dummy data in λ_{12} and lower trace to a signal in λ_{11}; (b) the third wave by another tunable laser was overlaid on the λ_{12} output arm (© 1991 IEEE).

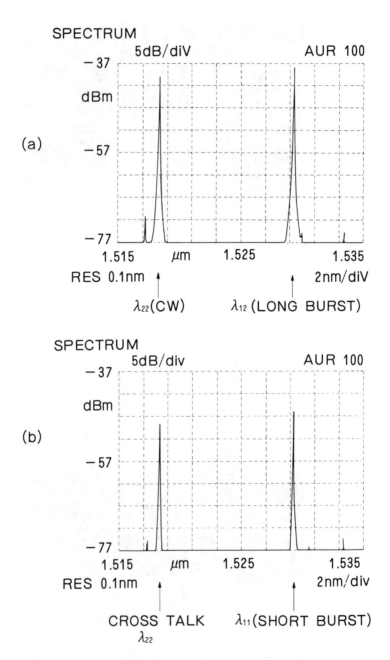

Figure 8.20 Measured spectrum of three waves by an optical spectrum analyzer; (a) and (b) correspond to two time-averaged outputs of a guided-wave demultiplexer. Resolution is 0.1 nm (© 1991 IEEE).

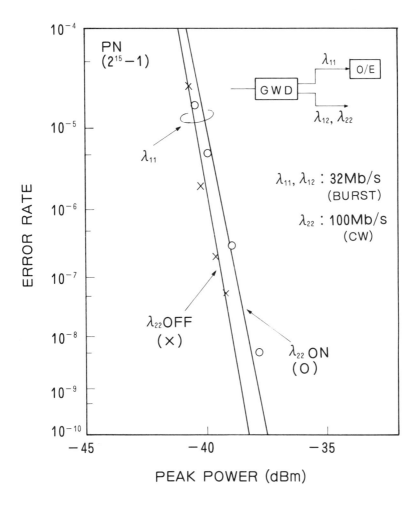

Figure 8.21 Measured error rate of a 32-Mb/s burst transmission as a function of peak power. Data with and without λ_{22} are plotted using circles and crosses, respectively (© 1991 IEEE).

The second experiment concerns three waves: λ_{11}, λ_{12}, and λ_{21}. By modifying the injection current into the Bragg region of a DBR-LD, the third wave was λ_{21} in this case. One subscriber received two waves: λ_{11} for 32-Mb/s burst transmission and λ_{21} for 100-Mb/s continuous transmission (i.e., not burst transmission). Two waves were separated by a WD. The measured error rates as a function of received average power for 100-Mb/s data (λ_{21}) are shown in Figure 8.22. Circles and crosses stand for data measured with and without λ_{11} and λ_{12}, respectively. There is no degradation from the upgrade. This is due to the large loss difference property between pass and rejection bands of a WD.

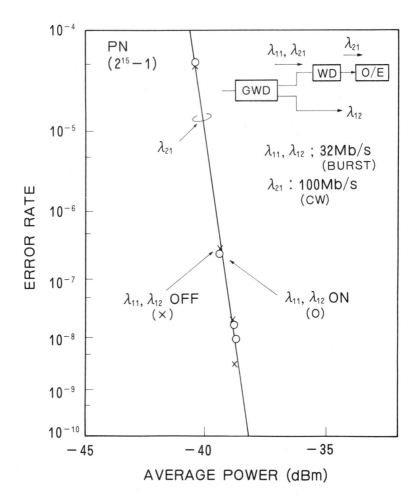

Figure 8.22 Measured error rate of a 100-Mb/s transmission as a function of average power. Data with and without λ_{11} and λ_{12} are plotted using circles and crosses, respectively (© 1991 IEEE).

8.4 MULTICHANNEL SYSTEM USING COHERENT TECHNOLOGY

8.4.1 Distribution Systems

The proposed and experimental distribution systems are shown in Figure 8.23, and star couplers are used as a combiner (a multiplexer) and as a splitter [31–37]. Lasers with slightly different optical frequencies are used as a light source, and the frequencies are locked by the absolute optical frequency control and channel space control. The laser light is modulated by IM or FSK. Optical amplification is used,

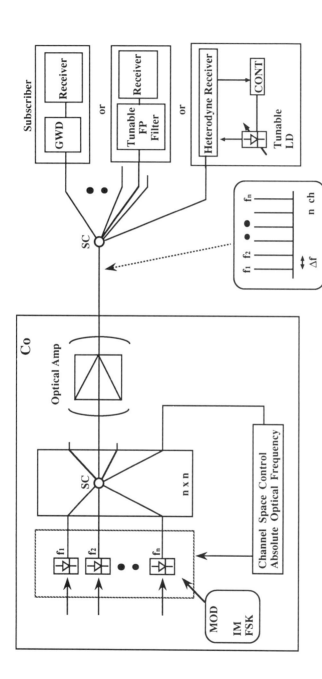

Figure 8.23 Distribution system using coherent transmission.

if required. Several detection schemes were proposed and demonstrated, such as heterodyne detection and direct detection with a tunable filter.

Direct detection for intensity modulated optical signals was used with an MZ-GWD as a tunable optical filter. A GWD selects one of the transmitted optical frequencies and the selected signals are directly detected by an ordinary receiver (see Figure 8.24(a)). Sixteen channels of 622 Mb/s with optical frequency spacing of $\Delta f = 5$ GHz were demonstrated [33].

Optical heterodyne detection for FSK-modulated signals was used, and the FSK #i signals f_i slightly deviate in an optical frequency domain according to the mark or space in FSK format, as shown in Figure 8.24(b). Here, frequency f_i in FSK stands for two frequencies of mark f_{i1} and space f_{i0}. Several experiments were reported [31,34,35], and one example is a 10-channel experiment of 400 Mb/s with $\Delta f = 8$ GHz [34].

FSK-modulated signals in OFDM systems can be selected and demodulated by a tunable FP filter with intensity detection (see Figure 8.25) [32], or by an MZ-GWD and an optical frequency discriminator with amplitude detection (see Figure 8.26) [36,37]. In the system of Figure 8.25, either f_{i1} (mark) or f_{i0} (space) is selected by an FP filter for f_i selection. This figure is drawn by assuming f_{i1} selection. In this case, the mark is determined by the detected light level above the threshold, and the space by the level below the threshold. For f_{i+1} selection, an FP filter is tuned to the frequency f_{i+1}. A four-channel experiment of 45 Mb/s with $\Delta f = 300 \sim 400$ MHz was demonstrated [32]. In the case of Figure 8.26, MZ-GWDs are used for both an optical frequency discriminator and an optical frequency selector. The selected FSK-modulated optical frequency f_i signals are converted to signals with amplitude changes by the combination of an MZ-GWD and balanced-mixer-type photodetection. The outputs of two arms of an MZ-GWD are changed according to the input optical frequency, as shown in Figure 8.26. Through the function of a balanced mixer, these outputs are subtracted and the frequency discriminator is realized. A 100-channel experiment of 622 Mb/s with $\Delta f = 10$ GHz was demonstrated with this method [36,37].

System parameters in typical proposed experimental OFDM multichannel distribution systems are summarized in Table 8.1. Many systems use FSK, and several demodulation schemes are proposed and tested. Up to 100 channels were demonstrated, showing the multichannel potential of OFDM systems. The optical frequency space Δf ranges from several hundreds of megahertz to 10 GHz, and the ratio of Δf to bit rate ranges from 7 to 55.

8.4.2 Bidirectional Systems

Several bidirectional systems are possible by combining the proposed technologies, such as heterodyne detection and direct detection with a tunable filter. One proposed

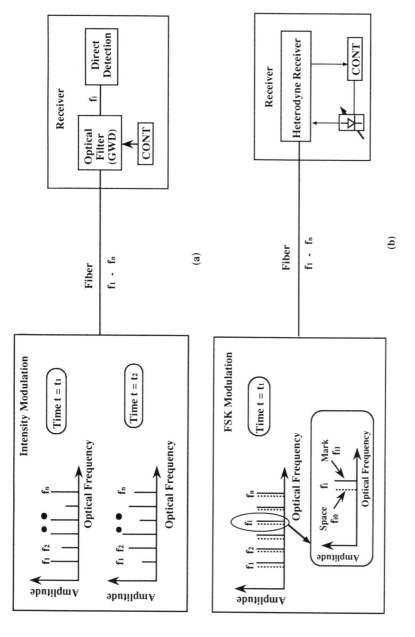

Figure 8.24 Modulation/demodulation in OFDM (1): (a) direct detection; (b) optical heterodyne detection.

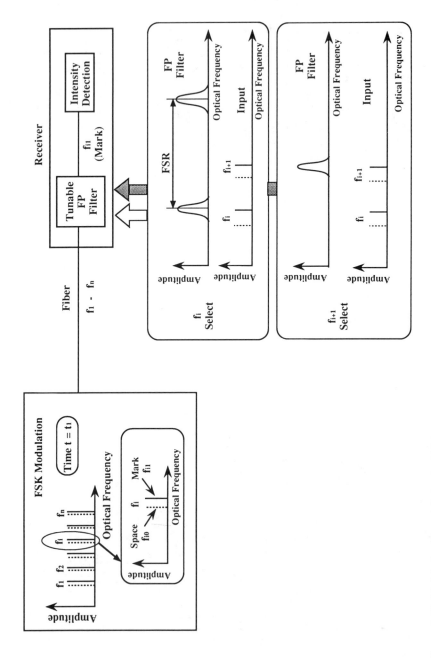

Figure 8.25 Modulation/demodulation in OFDM (2).

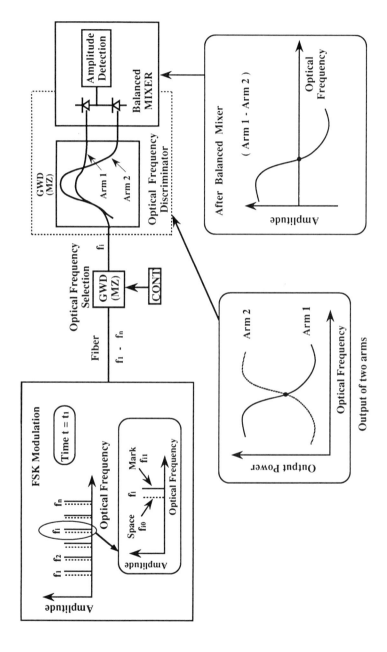

Figure 8.26 Modulation/demodulation in OFDM (3).

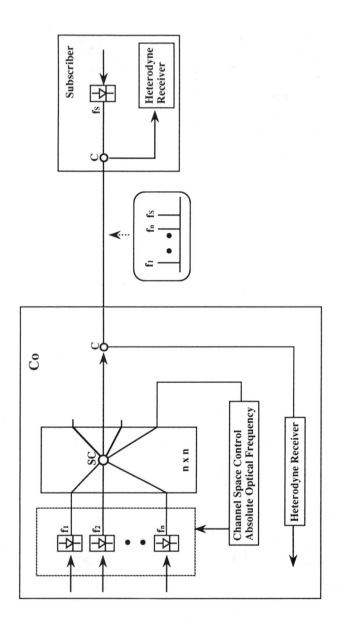

Figure 8.27 Bidirectional systems.

Table 8.1
Parameters in Experimental Coherent Technology–Based Distribution Systems

List Number	Optical Modulation	Demodulation	Channel Number	Δf	Bit rate	Reference
1	FSK	Optical heterodyne	4	300 MHz	45 Mb/s	[31]
2	FSK	Tunable FP filter Detection of intensity	4	300 ~ 400 MHz	45 Mb/s	[32]
3	IM	GWD, DD	16	5 GHz	622 Mb/s	[33]
4	FSK	Optical heterodyne	10	8 GHz	400 Mb/s	[34]
5	FSK	Optical heterodyne	16	8.5 GHz	155 Mb/s	[35]
6	FSK	GWD, Optical discriminator detection of intensity	100	10 GHz	622 Mb/s	[36] [37]

system is shown in Figure 8.27 [38,39]. This system uses a heterodyne detection scheme for both direction transmissions. SC an C stand for a star coupler and a coupler, respectively. Optical frequencies $f_1 \sim f_n$ are used for downstream transmission and f_s is used for upstream transmission.

REFERENCES

[1] Nash, F. R., "Mode Guidance Parallel to the Injection Plane of Double-Heterostructure GaAs Lasers," *J. Appl. Phys.*, Vol. 44, 1973, p. 4696.
[2] Okuda, M., and K. Onaka, "Tunability of Distributed Bragg-Reflector Laser by Modulating Refractive Index in Corrugated Waveguide," *Japan. J. Appl. Phys.*, Vol. 16, 1977, p. 1501.
[3] Tohmori, Y., Y. Suematsu, H. Tsushima, and S. Arai, "Wavelength Tuning of GaInAsP/InP Integrated Laser With Butt-Jointed Built-in Distributed Bragg Reflector," *Electron. Lett.*, Vol. 19, 1983, p. 656.
[4] Westbrook, L. D., A. W. Nelson, P. J. Fiddyment, and J. B. Collins, "Monolithic 1.5μm Hybrid DFB/DBR Lasers With 5 nm Tuning Range," *Electron. Lett.*, Vol. 20, 1984, p. 957.
[5] Murata, S., I. Mito, and K. Kobayashi, "Over 720 GHz (5.8 nm) Frequency Tuning by a 1.5 μm DBR Laser With Phase and Bragg Wavelength Control Regions," *Electron. Lett.*, Vol. 23, 1987, p. 403.
[6] Nilsson, O., and Y. Yamamoto, "Small-Signal Response of a Semiconductor Laser With Inhomogeneous Linewidth Enhancement Factor: Possibility of a Flat Carrier-Induced FM Response," *Appl. Phys. Lett.*, Vol. 46, 1985, p. 223.
[7] Yoshikuni, Y., and G. Motosugi, "Independent Modulation in Amplitude and Frequency Regimes by a Multielectrode Distributed Feedback Laser," *Conf. on Optical Fiber Communication '86 (OFC'86)*, 1986, p. 32.
[8] Yoshikuni, Y., K. Oe, G. Motosugi, and T. Matuoka, "Broad Wavelength Tuning Under Single-Mode Oscillation With a Multielectrode Distributed Feedback Laser," *Electron. Lett.*, Vol. 22, 1986, p. 1153.

[9] Kuznetsov, M., "Theory of Wavelength Tuning in Two-Segment Distributed Feedback Lasers," *IEEE J. Quantum Electron.*, Vol. 24, 1988, p. 1837.

[10] Numai, T., S. Murata, and I. Mito, "1.5 μm Wavelength Tunable Phase-Shift Controlled Distributed Feedback Laser Diode With Constant Spectral Linewidth in Tuning Operation," *Electron. Lett.*, Vol. 24, 1988, p. 1526.

[11] Yoshikuni, Y., and G. Motosugi, "Multielectrode Distributed Feedback Laser for Pure Frequency Modulation and Chirping Suppressed Amplitude Modulation," *IEEE J. of Lightwave Technol.*, Vol. LT-5, 1987, p. 516.

[12] Amann, M. C., S. Illek, C. Schanen, W. Thulke, and H. Lang, "Continuously Tunable Single-Frequency Laser Diode Utilizing Transverse Tuning Scheme," *Electron. Lett.*, Vol. 25, 1989, p. 837.

[13] Wyatt, R., and W. J. Devlin, "10 kHz Linewidth 1.5 μm InGaAsP External Cavity Laser With 55 nm Tuning Range," *Electron. Lett.*, Vol. 19, 1983, p. 110.

[14] Bagley, M., R. Wyatt, D. J. Elton, H. J. Wickes, P. C. Spurdens, C. P. Seltzer, D. M. Cooper, and W. J. Devlin, "242 nm Continuous Tuning From a GRIN-SC-MQW-BH InGaAsP Laser in an Extended Cavity," *Electron. Lett.*, Vol. 26, 1990, p. 267.

[15] Coguin, G., K. W. Cheung, and M. M. Choy, "Single- and Multiple-Wavelengh Operation of Accousto-Optically Tuned Semiconductor Lasers at 1.3 μm," *Proc. 11th IEEE International Semiconductor Laser Conf.*, Boston, 1988, p. 130.

[16] Yariv, A., *Quantum Electronics*, 2nd ed., John Wiley & Sons, 1975.

[17] Heismann, F., R. C. Alferness, L. L. Buhl, G. Eisenstein, S. K. Korotky, J. J. Veselka, L. W. Stulz, and C. A. Burrus, "Narrow-Linewidth Electro-Optically Tunable InGaAsP-T_i: L_iNbO_3 Extended Cavity Laser," *Appl. Phys. Lett.*, Vol. 51, 1987, p. 164.

[18] Tang, W. T., N. A. Olsson, R. A. Linke, and R. A. Logan, "1.5 μm Wavelength GaInAsP C^3 Lasers: Single-Frequency Operation and Wideband Frequency Tuning," *Electron. Lett.*, Vol. 19, 1983, p. 415.

[19] Kobayashi, K., and I. Mito, "Single Frequency and Tunable Laser Diodes," *IEEE J. of Lightwave Technol.*, Vol. 6, 1988, p. 1623.

[20] Takato, N., K. Inoue, K. Jinguji, M. Yasu, H. Toba, and M. Kawachi, "Guided-Wave Multi/Demultiplexer for Optical FDM Transmission," *Proc. 12 th European Conference of Optical Communication (ECOC '86)*.

[21] Inoue, K., N. Takato, H. Toba, and M. Kawachi, "A Four-Channel Optical Waveguide Multi/Demultiplex for 5 GHz Spaced Optical FDM Transmission," *IEEE J. of Lightwave Technol.*, Vol. 6, 1988, p. 339.

[22] Marcatilli, E. A. J., "Dielectric Rectangular Waveguide and Directional Coupler for Integrated Optics," *B.S.T.J.*, Vol. 48, 1969, p. 2071.

[23] Kobayashi, S., N. Shibata, S. Shibata, and T. Izawa, "Characteristics of Optical Fibers in the Infrared Waveguide Region," *Rev. Elect. Commun. Lab. (Japan)*, Vol. 26, 1978, p. 453.

[24] Kashima, N., "Upgrade of Passive Optical Subscriber Network," *IEEE J. of Lightwave Technol.*, Vol. 9, 1991, p. 113.

[25] Toba, H., K. Oda, N. Takato, and K. Nosu, "5-GHz Spaced, Eight-Channel, Guided-Wave Tunable Multi/Demultiplexer for Optical FDM Transmission Systems," *Electron. Lett.*, Vol. 23, 1987, p. 788.

[26] Saleh, A. A. M., and J. Stone, "Two-Stage Fabry-Perot Filters as Demultiplexers in Optical FDMA LAN's," *IEEE J. of Lightwave Technol.*, Vol. 7, 1989, p. 323.

[27] Kawaguchi, H., K. Magari, K. Oe, Y. Noguchi, Y. Nakano, and G. Motosugi, "Optical Frequency-Selective Amplification in a Distributed Feedback Type Semiconductor Laser Amplifier," *Appl. Phys. Lett.*, Vol. 50, 1987, p. 66.

[28] Numai, T., S. Murata, and I. Mito, "1.5 μm Tunable Wavelength Filter Using a Phase-Shift-Controlled Distributed Feedback Laser Diode With a Wide Tuning Range and a High Constant Gain," *Appl. Phys. Lett.*, Vol. 54, 1989, p. 1859.

[29] Chraplyvy, A. R., and R. W. Tkach," Narrowband Tunable Optical Filter for Channel Selection in Densely Packed WDM Systems," *Electron. Lett.*, Vol. 22, 1986, p. 1084.
[30] Kashima, N., and K. Kikushima, "New Optical Star-Bus Network for Subscriber," *J. Opt. Commun.*, Vol. 11, 1990, p. 42.
[31] Glance, B. S., J. Stone, K. J. Pollock, P. J. Fitzgerald, C. A. Burrus, Jr., B. L. Kasper, and L. W. Stulz, "Densely Spaced Coherent Star Network With Optical Signals Confined to Equally Spaced Frequencies," *IEEE J. of Lightwave Technol.*, Vol. 6, 1988, p. 1770.
[32] Kaminow, I. P., P. P. Iannone, J. Stone, and L. W. Stulz, "FDMA-FSK Star Network With a Tunable Optical Filter Demultiplexer," *IEEE J. of Lightwave Technol.*, Vol. 6, 1988, p. 1406.
[33] Toba, H., K. Oda, K. Nosu, and N. Takato, "16-Channel Optical FDM Distribution/Transmission Experiment Utilizing Multichannel Frequency Stabilizer and Waveguide Frequency Selection Switch." *Electron. Lett.*, Vol. 25, 1989, p. 574.
[34] Yamazaki, S., M. Shibutani, N. Shimosaka, S. Murata, T. Ono, M. Kitamura, K. Emura, and M. Shikada, "A Coherent Optical CATV Distribution System," *IEEE J. of Lightwave Technol.*, Vol. 8, 1990, p. 396.
[35] Welter, R., W. B. Sessa, M. W. Meada, R. E. Wagner, L. Curtis, J. Young, T. P. Lee, K. Kandruri, H. Kodera, Y. Koga, and J. R. Barry, "Sixteen-Channel Coherent Broadcast Network at 155 Mb/s," *IEEE J. of Lightwave Technol.*, Vol. 7, 1989, p. 1438.
[36] Toba, H., K. Oda, K. Nakanishi, N. Shibata, K. Nosu, N. Takato, and K. Sato, "100-Channel Optical FDM Transmission/Distribution at 622 Mb/s Over 50 km Utilizing a Waveguide Frequency Selection Switch," *Electron. Lett.*, Vol. 26, 1990, p. 376.
[37] Toba, H., K. Oda, K. Nakanishi, N. Shibata, K. Nosu, N. Takato, and M. Fukuda, "100- Channel Optical FDM Transmission/Distribution at 622 Mb/s Over 50 km," *IEEE J. of Lightwave Technol.*, Vol. 8, 1990, p. 1396.
[38] Bachus, E.-J., R. P. Braun, W. Eutin, E. Großmann, H. Foisel, K. Heimes, and B. Strebel, "Coherent Optical Fiber Subscriber Line," *Proc. of IOOC-ECOC'85*, 1985, p. 61.
[39] Bachus, E.-J., R. P. Braun, C. Caspar, H.-M. Foisel, E. Großmann, B. Strebel, and F.-J. Westphal, "Coherent Optical Multichannel Systems," *IEEE J. of Lightwave Technol.*, Vol. 7, 1989, p. 375.

Chapter 9
SCM Systems

SCM systems are attractive from the multichannel transmission viewpoint. In this chapter, we discuss multiple-access and multichannel video distribution systems using SCM. The EDFA is useful for increasing the number of subscribers in multichannel video distribution systems.

9.1 MULTIPLE-ACCESS NETWORK

Multiple-access networks using SCM and a star coupler have been proposed [1,2], and they are shown in Figure 9.1. In SCM, both analog and digital signals can be simultaneously transmitted. Although these figures are drawn by assuming direct intensity modulation, it is possible to use an external modulator.

The system shown in Figure 9.1(a) is intended for a network like LAN. Station #i receives the predetermined signals of an electric subcarrier f_i and sends any one of the signals of an electric subcarrier $f_n (n = 1 \sim N)$. To avoid a signal collision, a mechanism for access control is required. This is the realization of the multiple-access network in an electric domain, and it corresponds to the network shown in Figure 7.6(b) in an optical domain.

In the system shown in Figure 9.1(b), bidirectional signals are multiplexed by SCM, f_n ($n = 1 \sim N$) for an upstream transmission and $F_n(n = 1 \sim N)$ for a downstream transmission. Each subscriber (#i) sends and receives the unique signals carried by a subcarrier f_i and F_i. Limitations due to an optical beat interference as discussed in Section 7.3 may exist at the CO receiver for this system.

The receiver of these SCM systems need only have the bandwidth of the desired channel, while the receiver in TDM systems needs to have the total bandwidth for all channels. This results in the increase of receiver sensitivity [3]. The system using SCM requires one wavelength, which results in a cost-effective system when compared to a system using WDM. An electric subcarrier instead of an optical carrier (wavelength) is dedicated to each subscriber. Since each subscriber can receive all

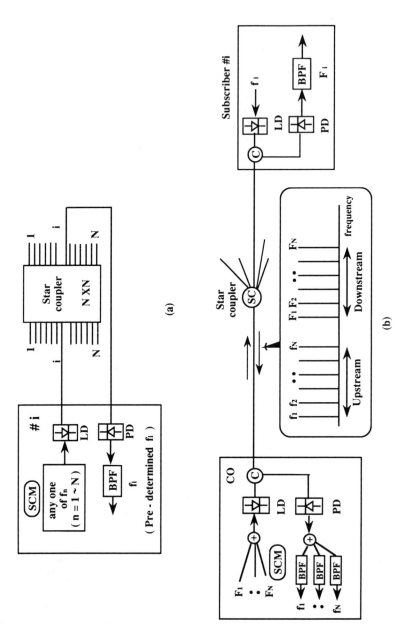

Figure 9.1 Multiple-access networks using SCM: (a) for a network; (b) for upstream and downstream transmission.

information by changing the BPF, these systems are less attractive from the privacy or security point of view.

9.2 MULTICHANNEL VIDEO DISTRIBUTION SYSTEM

Multichannel video distribution systems using the SCM system shown in Figure 9.1(b) can be realized without upstream signals [4,5]. For more channel transmissions, multichannel video distribution systems using the combination of SCM and WDM, and SCM and OFDM have been investigated [6,7], and they are shown in Figure 9.2. Since WDM or OFDM filters are located at the front of an optical receiver, only one wavelength (frequency) is received by an optical receiver. Therefore, there is no degradation due to optical beat interference. Several optical filters (discussed in Chapters 7 and 8) can be applied to these systems. A 160-channel FM video transmission ($n = 10$ and $N = 16$ in Figure 9.2) was reported by using 16 DFB lasers with 10-GHz spacing ($\Delta f = 10$ GHz) [7]. In the system, a Fabry-Perot etalon was used as an optical filter.

In the case of the analog video signal transmission, the distortion must be taken into consideration. When we use the direct intensity modulation of a laser diode, it is important to estimate the distortion due to the nonlinearity of LD modulation. The reported *intermodulation distortion* (IMD), *second harmonic distortion* (2HD), and *third harmonic distortion* (3HD) are nearly the same for a GaAlAs laser diode and an InGaAsP laser diode, and there are no significant differences between an FP-LD and a DFB-LD [8–10]. These distortions have a peak near the frequency range between $f_0/2$ and f_0, where f_0 is the relaxation response frequency of a laser diode. It was reported that the peak values of IMD, 2 HD, and 3 HD are about -25 dBc, about -10 dBc, and about -20 dBc, respectively [10]. The notation dBc stands for the electrical decibels relative to the carrier. For the frequency below $f_0/2$, these distortion decreases and the value of -40 dBc ~ -70 dBc were measured in the low-frequency range. When the modulation depth of a laser becomes small, the distortion tends to be small; however, the signal power also becomes small. It is useful to use a laser with higher f_0 value.

For the purpose of minimizing the distortion, it is common for LD analog modulation to set a bias current well above the threshold current, as discussed in Section 3.1. The shot noise increases due to this high-bias current, and this is one of the fundamental limits on channel number in SCM-based video distribution systems. A high C/N is required for AM video transmission, and this limitation is severe when compared to FM or digital transmission. According to the reported estimation, 50 channels per milliwatt of received optical power is possible for National Television System Committee (NTSC) vestigial sideband–AM (VSB-AM) (bandwidth 4 MHz) with C/N = 55 dB [11]. The use of an external modulator may be another approach to avoiding the distortions.

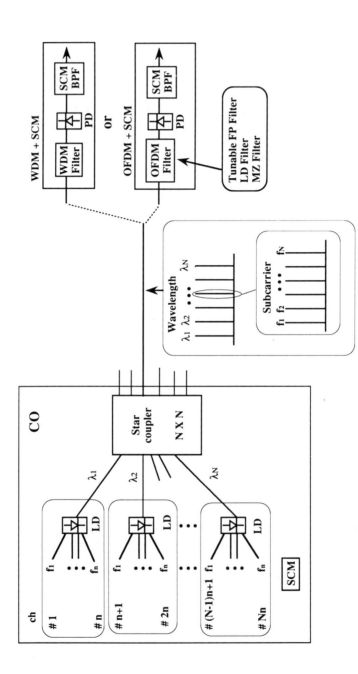

Figure 9.2 Multichannel video distribution systems (1).

Multichannel transmission systems using the combination of SCM and optical coherent heterodyne techniques have been investigated for improving receiver sensitivity [12]. For multichannel video distribution systems using SCM and coherent heterodyne techniques, the sharing of a local oscillator laser has been proposed; they are shown in Figure 9.3 [13,14]. By applying SCM for multichannel transmission, only one laser is used for a transmitter. Therefore, the sharing of a local oscillator laser is possible. In Figure 9.3(a), a local oscillator laser is located at the hub, and both coherently modulated optical signals and local oscillator laser light are mixed and distributed by a $2 \times N$ star coupler. At the subscriber, optical signals are demodulated in a heterodyne scheme and one channel from the SCM signals is selected by a video tuner. Another proposed system is shown in Figure 9.3(b), and a local oscillator laser is located at the CO. PM and FM modulations were used in the experiments of [13,14].

9.3 ENHANCED MULTICHANNEL VIDEO DISTRIBUTION SYSTEM USING EDFA

The EDFA has attractive features for many applications, as discussed in Chapter 5. It has low noise, high gain (intrinsic high gain and low fiber-connection loss), high saturation output power, and low distortion (good linearity). Long nonrepeated transmission experiments of 1.5-μm wavelength using the EDFA at a high bit rate have been successively demonstrated for a trunk system application [15–17]. The applications of the EDFA for multichannel video distribution systems using SCM are shown in Figure 9.4. The EDFA increases the distribution number by compensating for the star coupler loss (distribution loss). Postamplification and inline amplifications are shown in Figure 9.4 and the combination of postamplification and inline amplification is also possible. A 160-channel FM video transmission using the EDFA as a postamplifier with a net gain of about 9 dB has been reported [7]. A 19-channel AM-VSB (or 11-channel FM) video transmission using b-stage cascade EDFAs as inline amplifiers has been reported [18]. Maximum repeater gains as an inline amplifier were about 11 and 17 dB for AM and FM experiments, respectively [18]. In AM applications, they require a high C/N and low distortion. With these experiments, the EDFA has been proved to have superior characteristics not only for trunk applications but also for video distribution applications. When the EDFA is used, nonlinear effects in a fiber (discussed in Chapter 3) must be taken into consideration because of high optical power.

It is important to calculate the noise in distribution systems with the EDFA, not only in SCM systems but also in more general distribution systems. The model for noise calculation in general is shown in Figure 9.5. A distribution network, which includes star couplers, fiber cables, and splices, is used between a receiver and the EDFA. The loss of the distribution network is assumed to be L. The gain, the population inversion parameter, and the optical bandwidth of the EDFA are assumed to

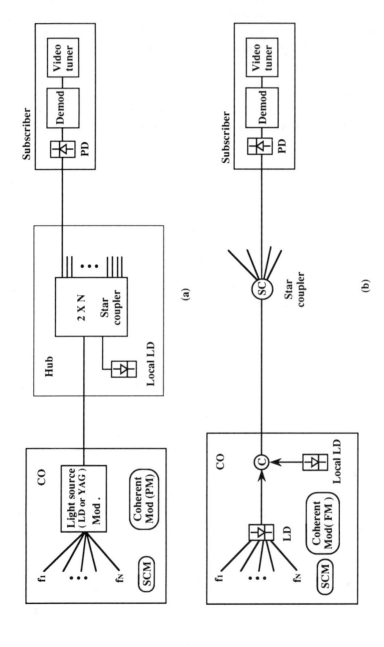

Figure 9.3 Multichannel video distribution systems (2): (a) oscillator located at the hub; (b) oscillator located at the CO.

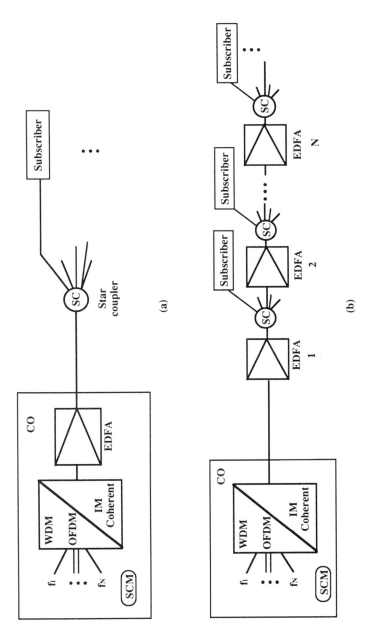

Figure 9.4 Multichannel video distribution systems using EDFA: (a) postamp; (b) inline amp.

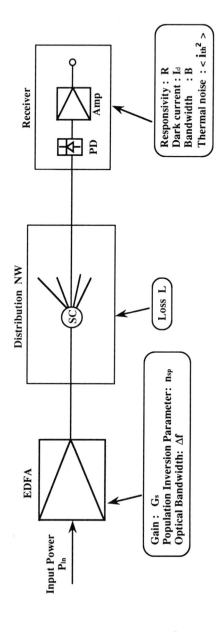

Figure 9.5 Model for noise calculation in distribution systems.

be G_s, n_{sp}, and Δf, respectively. A photodiode in the receiver has the responsivity R and the dark current I_d. The bandwidth and thermal noise of the receiver are assumed to be B and $\langle i_{th}^2 \rangle$, respectively. For calculations, the equations of (5.47) and (5.48) can be used with slight modifications. Here, we assume a coherent light. The input optical power P_{in} of the EDFA is related to $\langle n_{in} \rangle$, which is the average input photon number, by the following equation:

$$P_{in} = hf \langle n_{in} \rangle \tag{9.1}$$

The received signal power $\langle i \rangle^2$ is

$$\langle i \rangle^2 = (R\, G_s\, L\, P_{in})^2 \tag{9.2}$$

and the total noise $\langle i^2 \rangle$ is the sum of the following noises:

$$\text{signal shot noise} = 2eRG_sLP_{in}B;$$
$$\text{spontaneous shot noise} = 4eRhf(G_s - 1)Ln_{sp}\Delta fB;$$
$$\text{PD dark current shot noise} = 2eI_dB;$$
$$\text{signal-spontaneous beat noise} = 4R^2hfG_s(G_s - 1)L^2n_{sp}P_{in}B;$$
$$\text{spontaneous-spontaneous beat noise} = 4(Rhf)^2(G_s - 1)^2L^2n_{sp}^2\Delta fB; \text{ and}$$
$$\text{thermal noise} = \langle i_{th}^2 \rangle.$$

and the result is

$$\langle i^2 \rangle = 2eRG_s\, L\, P_{in}\, B + 4eRhf\, (G_s - 1)\, L\, n_{sp}\, \Delta f\, B$$
$$+ 2e\, I_d\, B + 4R^2\, hfG_s(G_s - 1)\, L^2\, n_{sp}\, P_{in}\, B \tag{9.3}$$
$$+ 4\, (Rhf)^2(G_s - 1)^2\, L^2\, n_{sp}^2\, \Delta f\, B + \langle i_{th}^2 \rangle$$

These equations can be rewritten by using quantum efficiency η instead of R through the equation $R = e\eta/(hf)$.

REFERENCES

[1] Darcie, T. E., "Subcarrier Multiplexing for Multiple-Access Lightwave Networks," *IEEE J. of Lightwave Technol.*, Vol. LT-5, 1987, p. 1103.
[2] Darcie, T. E., P. P. Ianonne, B. L. Lasper, J. R. Talman, C. A. Burrus, and T. A. Baker, "Wide-Band Lightwave Distribution System Using Subcarrier Multiplexing," *IEEE J. of Lightwave Technol.*, Vol. 7, 1989, p. 997.
[3] Darcie, T. E., M. E. Dixon, B. L. Kasper, and C. A. Burrus, "Lightwave System Using Microwave Subcarrier Multiplexing," *Electron. Lett.*, Vol. 22, 1986, p. 774.
[4] Olshansky, R., V. A. Lanzisera, and P. M. Hill, "Subcarrier Multiplexed Lightwave Systems for Broad-Band Distribution," *IEEE J. of Lightwave Technol.*, Vol. 7, 1989, p. 1329.

[5] Way, W. I., "Subcarrier Multiplexed Lightwave System Design Consideration for Subscriber Loop Applications," *IEEE J. of Lightwave Technol.*, Vol. 7, 1989, p. 1806.

[6] Westlake, H. J., G. R. Hill, G. E. Wickens, and B. P. Cavanagh, "Subcarrier Multiplexed Transmission Using Wavelength Division Multiplexing and Optical Amplifier," *Electron. Lett.*, Vol. 25, 1989, p. 632.

[7] Way, W. I., M. W. Maeda, A. Y. Yan, M. J. Andrejco, M. M. Choy, M. Saifi, and C. Lin, "160-Channel FM-Video Transmission Using Optical FM/FDM and Subcarrier Multiplexing and an Erbium Doped Optical Fiber Amplifier," *Electron. Lett.*, Vol. 26, 1990, p. 139.

[8] Strubkjaer, K., and M. Danielsen, "Nonlinearity of GaAlAs Lasers—Harmonic Distortion," *IEEE J. of Quantum Electron.*, Vol. QE-16, 1980, p. 531.

[9] Hong, T., T. Suematu, S. Chung, and M. Kang, "Harmonic Characteristics of Laser Diodes," *J. Opt. Commun.*, Vol. 3, 1982, p. 42.

[10] Darcie, T. E., R. S. Tucker, and G. J. Sullivan, "Intermodulation and Harmonic Distortion in InGaAs Lasers," *Electron. Lett.*, Vol. 21, 1985, p. 665.

[11] Saleh, A. A. M., "Fundamental Limit on Number of Channels in Subcarrier-Multiplexed Lightwave CATV System," *Electron. Lett.*, Vol. 25, 1989, p. 776.

[12] Gross, R., and R. Olshansky, "Multichannel Coherent FSK Experiments Using Subcarrier Multiplexing Techniques," *IEEE J. of Lightwave Technol.*, Vol. 8, 1990, p. 406.

[13] Gross, R., W. Rideout, R. Olshansky, and G. R. Joyce "Heterodyne Video Distribution Systems Sharing Transmitter and Local Oscillator Lasers," *IEEE J. of Lightwave Technol.*, Vol. 9, 1991, p. 524.

[14] Watanabe, S., I. Yokota, T. Naito, T. Chikama, H. Kuwahara, and T. Touge, "Polarization-Insensitive Optical Coherent SCM System Using Mixed Lightwave Transmission of Signal and Local Laser Light," *Electron. Lett.*, Vol. 27, 1991, p. 361.

[15] Hagimoto, K., K. Iwatsuki, A. Takada, M. Nakazawa, M. Saruwatari, A. Aida, K. Nakagawa, and M. Horiguchi, "250 km Nonrepeated Transmission at 1.8 Gb/s Using LD Pumped Er^{3+}-Doped Fibre Amplifiers in IM/Direct Detection System," *Electron. Lett.*, Vol. 25, p. 662.

[16] Edagawa, N., K. Mochizuki, and H. Wakabayashi, "1.2 Gbit/s, 218 km Transmission Experiment Using Inline Er-Doped Optical Fibre Amplifiers," *Electron. Lett.*, Vol. 25, 1989, p. 363.

[17] Taga, H., Y. Yoshida, N. Edagawa, S. Yamamoto, and H. Wakabayashi, "459 km, 2.4 Gbit/s Four Wavelength Multiplexing Optical Fibre Transmission Experiment Using Six Er-Doped Fibre Amplifiers," *Electron. Lett.*, Vol. 26, 1990, p. 500.

[18] Kikushima, K., E. Yoneda, and K. Aoyama, "6-Stage Cascade Erbium Fiber Amplifiers for Analog AM- and FM-FDM Video Distribution Systems," *Proc. of OFC'90*, PD22, 1990.

Chapter 10
Future Subscriber Systems

Optical subscriber loop systems are on the way to being constructed, while optical trunk transmission systems have been deployed worldwide. Several approaches have been tried, and the effort of realizing optical subscriber loops will continue. In this chapter, future optical subscriber loop systems and devices are discussed. These systems may be only dreams or science fiction at present; however, some of them will be a reality in the future.

10.1 PHOTONIC INTEGRATED CIRCUITS

Electric ICs and LSIs are a great success. They have turned complex electric circuits into small chips, and their cost is becoming cheaper and cheaper. Because of ICs and LSIs, digital transmission is widely used today. Optical devices are made by discrete components at present. Figure 10.1 is a schematic example of a transceiver. Although electric circuits such as LD drivers, receivers, and digital processing are made by LSI, optical devices such as LDs, PDs, and couplers in a transceiver are not integrated. To realize a low-cost transceiver, photonic integration must be developed, with which active optical devices (LD, PD) and passive waveguide devices (couplers) are integrated into one chip. Finally, photonic and electric integration will be commercially available. Developments towards this goal are now under study [1].

10.2 BROADBAND SYSTEMS USING OPTICAL SIGNAL PROCESSING

Optical signal processing is considered to be a promising technology for broadband signal processing to overcome the limitation of electrical signal processing [2–4]. When a sophisticated optical subscriber loop system is constructed, where broadband signals will be transmitted in the future, optical signal processing will be applied to the system. Optical amplifier, optical switching, optical memory, and wavelength conversion technologies have been investigated. Steps for the possible introduction

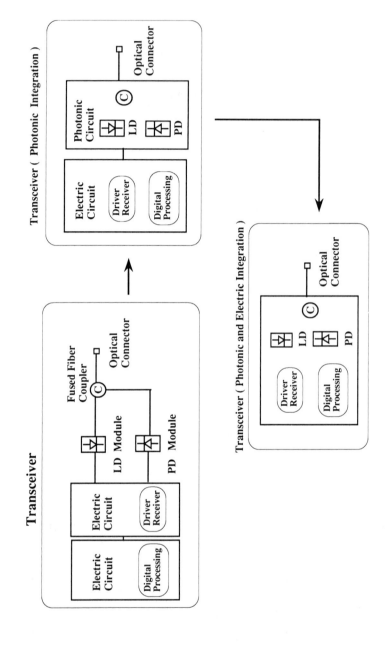

Figure 10.1 Photonic and electric integration of transceiver.

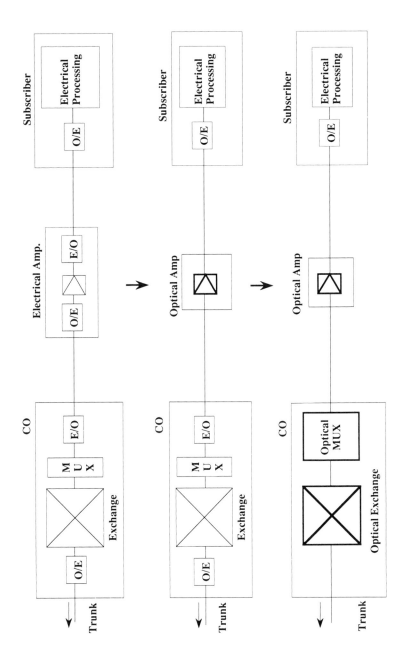

Figure 10.2 Introduction of optical processing.

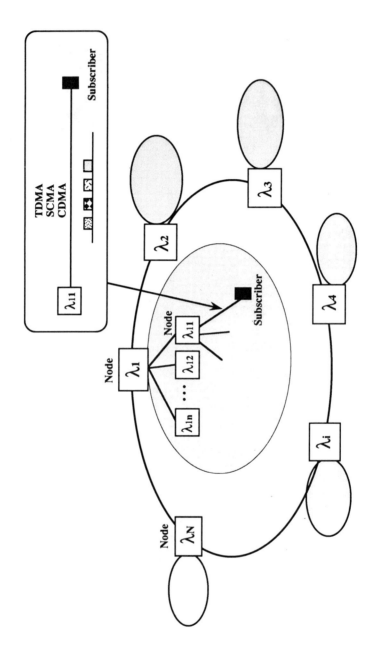

Figure 10.3 Multiwavelength network.

of optical processing technologies into subscriber loops are shown in Figure 10.2. Light emitting (LD), light transmitting (fiber), and light receiving (PD) belong to optical devices in present systems. When we need to amplify optical signals, they must be converted to electrical signals, and then they are again converted to the optical format. First, optical amplifiers will be used. Optical amplification has several advantages, such as bit rate and modulation format independence. Then optical exchange and multiplexing using optical switching and memory technologies will follow. At this point, electrical signal processing will remain only at subscriber transceivers, and broadband optical networks will be realized.

10.3 MULTIWAVELENGTH SYSTEM USING MULTIPLE-ACCESS METHOD

OFDM is a promising scheme for the multiple-access method. A single-mode fiber has a low-loss wavelength region from 1.2 ~ 1.7 μm, which corresponds to about 60 THz. The ratio of optical frequency space Δf to bit rate ranges from 7 to 55, as discussed in Chapter 8. Assuming this ratio to be 30, Δf equals 300 GHz for 10-Gb/s transmission. In this case, the number of available wavelengths is 200 (= 60 THz/300 GHz). Since we assume the ratio to be 30, the frequency utilization efficiency is about 0.03 (bits/s)/Hz. OFDM is superior in frequency utilization efficiency when compared to WDM [5,6]. Future subscriber systems using many wavelengths have been discussed. The system in Figure 10.3 is based on the concept of [7]. Wavelengths are uniquely dedicated to the node, and optical signals are routed according to their wavelengths. For example, the signals at node λ_{11}, whose destination is λ_i, are carried by light with wavelength λ_i. Since the number of wavelengths is around 200, the help of other multiplexing methods such as TDMA, SCMA, and CDMA will be necessary in this system.

REFERENCES

[1] Horimatu, T., and M. Sasaki, "OEIC Technology and Its Application to Subscriber Loops," *IEEE J. of Lightwave Technol.*, Vol. 7, 1989, p. 1612.
[2] Kazovsky, L. G., "Optical Signal Processing for Lightwave Communications Networks," *IEEE J. of Lightwave Technol.*, Vol. 8, 1990, p. 973.
[3] Fujimoto, N., H. Rokugawa, K. Yamaguchi, S. Masuda, and S. Yamakoshi," Photonic Highway: Broad-Band Ring Subscriber Loops Using Optical Signal Processing," *IEEE J. of Lightwave Technol.*, Vol. 7, 1989, p. 1798.
[4] Fujiwara, M., N. Shimosaka, M. Nishio, S. Suzuki, S. Yamazaki, S. Murata, and K. Kaede," A Coherent Photonic Wavelength-Division Switching System for Broad-Band Networks," *IEEE J. of Lightwave Technol.*, Vol. 8, 1990, p. 416.

[5] Nosu, K., and K. Iwashita, "A Consideration of Factors Affecting Future Coherent Lightwave Communication Systems," *IEEE J. of Lightwave Technol.*, Vol. 6, 1988, p. 686.
[6] Brackett, C. A., "Dense Wavelength Division Multiplexing Networks: Principles and Applications," *IEEE J. of Lightwave Technol.*, Vol. 8, 1990, p. 948.
[7] Cochrane, P., and M. Brain, "Future Optical Fiber Transmission Technology and Networks," *IEEE Communications Magazine*, November 1988, p. 45.

Appendix A
List of Acronyms

ADS	active doubler star
AGC	automatic gain control
AO	acousto-optical
APC	automatic power control
APD	avalanche photodiode
APSK	amplitude phase-shift keying
AR	antireflection
ASE	amplified spontaneous emission
ASK	amplitude shift keying
ATC	automatic temperature control
ATM	asynchronous transfer mode
BISDN	Broadband Integrated Services Digital Network
BPF	band-pass filter
C/N	carrier-to-noise ratio
CCIR	International Radio Consultative Committee
CCITT	Consultative Committee in International Telegraphy and Telephony
CDM	code-division multiplexing
CDMA	code-division multiple access
CIP	carrier-induced phase modulation
CMI	coded mark inversion
CO	central office
COMP	comparator
CPFSK	continuous phase FSK
CRC	cyclic redundancy check
CW	continuous wave
dBc	decibels relative to the carrier
DBR-LD	distributed Bragg reflector laser diode
DDM	directional division multiplexing
Demod.	demodulation

DEMUX	demultiplexer
DFA	doped-fiber amplifier
DFB-LD	distributed-feedback laser diode
DH LD	double heterostructure laser diode
DPSK	differential phase-shift keying
DSF	dispersion-sifted fiber
DSM laser	dynamic single-mode laser
EDFA	Er-doped-fiber amplifier
EQL	equalizer
FDMA	frequency-division multiple access
FET	field-effect transistor
FIL	filter
FP	Fabry-Perot
FP-LD	Fabry-Perot laser diode
FPA	Fabry-Perot amplifier
FR	frame synchronous bit
FSK	frequency shift keying
FTTC	fiber-to-the-curve
FTTH	fiber-to-the-home
FTTO	fiber-to-the-office
FWM	four-wave mixing
GMSK	Gaussian-filtered MSK
GVD	group velocity dispersion
GWD	guided-wave demultiplexer
HWHM	half width at half maximum
HZ design	high-impedance design
IC	integrated circuit
ID	identification
IF	intermediate frequency
IM	intensity modulation
IMD	intermodulation distortion
IM/DD	intensity modulation/direct detection
IMP	intermodulation product
IR	infrared
ISDN	integrated services digital network
JFET	junction FET
LAN	local-area network
LATT	laser attenuator
LD	laser diode
LD transceiver	laser transceiver
LED	light-emitting diode
LPF	low-pass filter

LSI	large-scale integrated circuit
LZ design	low-impedance design
MAN	metropolitan-area network
MCVD	modified chemical vapor deposition
MQW	multiple-quantum (or multiquantum) well
MSK	minimum shift keying
MUX	multiplexer
MZ	Mach-Zehnder
O/E	optical/electrical
NDFWM	nearly degenerate four-wave mixing
NISDN	narrow band integrated services digital network
NRZ	nonreturn to zero
NTSC	National Television System Committee
OFDM	optical frequency-division multiplexing
OFDMA	optical frequency-division multiple access
OH	overhead bit
OMI	optical modulation index
OPLL	optical PLL
OTDR	optical time domain reflectometer
OVD	outer vapor deposition
PC	polarization controller
PCM	pulse-code modulation
PD	photodiode
PDS	passive double star
PLL	phase-locked loop
PM	phase modulation
PN	pseudorandom noise
PON	passive optical network
PR	preamble bit
PSK	phase-shift keying
PTS	passive triple star
QAM	quadrature amplitude modulation
RC	resistance-capacitance
RIN	relative intensity noise
RT	remote terminal
RX	receiver
RZ	return to zero
S/N	signal-to-noise ratio
SBS	stimulated Brillouin scattering
SC	star coupler
SCM	subcarrier multiplexing
SCMA	subcarrier multiple access

SDM	space-division multiplexing
SDMA	space-division multiple access
SI fiber	step-index fiber
SIR	signal-to-interference ratio
SLDA	semiconductor laser diode amplifier
SPM	self-phase modulation
SRS	stimulated Raman scattering
SS	single star
STM	synchronous transfer mode
TCM	time-compression multiplexing
TCM CONT	TCM controller
TDMA	time-division multiple access
TE	transverse electric
TEM	transverse electromagnetic
TM	transverse magnetic
TO	telephone office
TTG laser	tunable twin-guide laser
TWA	traveling wave amplifier
TX	transmitter
TZ design	transimpedance design
UV	ultraviolet
VAD	vapor-phase axial deposition
VSB	vestigial sideband
WD	wavelength demultiplexer
WDM	wavelength-division multiplex (or multiplexing)
WDMA	wavelength-division multiple access
WKB	Wentzel-Kramers-Brillouin
WSSC	wavelength-selective star coupler
2 HD	second harmonic distortion
3 HD	third harmonic distortion

Index

Absorption, 32–33, 59
Access methods, 19–26
Acousto-optical devices, 9
Active double star network, 9
Active feedback, 140–42
Active layer thickness, 207, 216
ADS. *See* Active double star network
AGC. *See* Automatic gain control
Amplification, 4, 186–89
  See also Optical amplifiers
Amplified spontaneous emission, 174
Amplitude phase shift keying, 89
Amplitude shift keying, 89, 101, 103, 109, 122
Analog transmission, 13, 41–42, 54, 73, 76, 87–88, 109, 129, 289, 291
Antireflective coating, 169
AO devices. *See* Acousto-optical devices
APC. *See* Automatic power control
APD. *See* Avalanche photodiode
APSK. *See* Amplitude phase shift keying
AR coating. *See* Antireflective coating
Array connector, 44–45
ASE. *See* Amplified spontaneous emission
ASK. *See* Amplitude shift keying
Asynchronous demodulation, 99–102
Asynchronous detection, 89
Asynchronous transfer mode, 26–28
ATC. *See* Automatic temperature control
ATM. *See* Asynchronous transfer mode
Attenuation, 211–18
Automatic gain control, 129, 150–51
Automatic power control, 77–78
Automatic temperature control, 78
Avalanche photodiode, 57, 59, 64, 84–85, 150, 152

Balanced-mixer detection, 94–95
Band-pass filter, 16
Bias current, 75–76, 169, 291
Bias point, 54
Bidirectional transmission, 11–19, 229, 280–85, 289–90
BISDN. *See* Broadband integrated services digital network
Bit pattern independence, 132
Bit rate, 7, 13, 28, 35, 132, 210, 224–25, 235, 237–38
BPF. *See* Band-pass filter
Bragg reflection, 53, 248, 250
Brillouin scattering. *See* Stimulated Brillouin scattering
Broadband integrated services digital network, 28
Broadband systems, 299–303
Bulk-type WDM device, 68–70, 233
Burst data, 12–13, 19, 22, 198
Bus topology, 9

Capacitance, 15, 57, 61, 63, 209
Carrier effect, 56
Carrier-induced phase modulation, 115
Carrier-to-noise ratio, 88, 114–15
CCITT. *See* Consultative Committee in International Telegraphy and Telephony
CDM. *See* Code-division multiplexing
CDMA. *See* Code-division multiple access
Chirping, 57
CIP. *See* Carrier-induced phase modulation
Cladding, 29–30, 34
CMI. *See* Coded mark inversion
CMOS. *See* Complementary metal oxide semiconductor

309

C/N. *See* Carrier-to-noise ratio
Code-division multiple access, 19, 25–26
Code-division multiplexing, 11–13, 16–17, 19
Coded mark inversion, 15, 79, 132
Coherent transmission, 41–42, 54, 73, 106, 109, 111, 114–15, 122, 129, 145
   multichannel system, 278–85
   OFDM, 105–8
   principle of, 88–94
   receiver configuration, 94–105
   in subscriber loops, 108–9
   See also Optical frequency-division multiplexing
COMP. *See* Comparator
Comparator, 97
Complementary metal oxide semiconductor, 7
Connectors, 41–46
Consultative Committee in International Telegraphy and Telephony, 2, 27
Continuous phase FSK, 89
Continuous wave, 29
Core, 29–30
Corrugation, 53
Coupling loss, 65–68, 177, 200, 202, 233–34
   See also Fiber loss
CPFSK. *See* Continuous phase FSK
CRC bits. *See* Cyclic redundancy check bits
Crosstalk, 186, 235–36
CW. *See* Continuous wave
Cyclic redundancy check bits, 195

Dark current, 208–9
DBR-LD. *See* Distributed Bragg reflector laser diode
DD. *See* Direct detection
DDM. *See* Directional division multiplexing
Degradation, 152–53
Demodulation, 94, 97–100, 280–83
Demultiplexer, 253–65
DEMUX. *See* Demultiplexer
Detection, 106, 150, 280
   See also Asynchronous detection; Direct detection; Heterodyne detection; Homodyne detection; Photodetection; Synchronous detection
DFB-LD. *See* Distributed-feedback laser diodes
DH LD. *See* Double heterostructure laser diodes
Differential phase-shift keying, 89, 100
Digital transmission, 4, 26–28, 54, 73, 76, 86–87, 89, 91, 109, 129–30, 132, 289
Direct detection, 73–74, 79–88, 109–10, 112–14
Directional division multiplexing, 11–13, 144
Dispersion, 34–36, 53, 68–69, 123
Dispersion-sifted fibers, 35

Distributed Bragg reflector laser diode, 48, 51, 53, 261
Distributed-feedback laser diodes, 3, 50, 53
Distribution systems, 278–80
Dopants, 29
Doped-fiber amplifier, 155–57, 172–77
   See also Er-doped-fiber amplifier
Double heterostructure laser diodes, 168
Downstream transmission, 13, 16
DPSK. *See* Differential phase-shift keying
Dropping, 9
DSF. *See* Dispersion-sifted fibers
DSM. *See* Dynamic single-mode laser
Dynamic range, 149–50, 215–18
Dynamic single-mode laser, 53

EDFA. *See* Er-doped-fiber amplifier
Envelope detection, 89, 99
EQL. *See* Equalizer
Equalizer, 129, 137
Equivalent circuit, 57–58, 61–62
Er-doped-fiber amplifier, 87, 155–56, 160, 172–77, 183, 293–97
Error rate, 79–80, 82–83, 87, 94–105, 221, 277
External cavity tunable lasers, 251
Extinction ratio, 84, 259–61

Fabry-Perot amplifier, 155–56, 158, 160, 165, 181–83, 188, 218
Fabry-Perot interferometer, 89, 263–64
Fabry-Perot laser diode, 48–49, 52–53
Fabry-Perot resonance model, 161–62
FDMA. *See* Frequency-division multiple access
Feedback designs, 138–144
FET. *See* Field-effect transistor
Fiber couplers, 65–68
   See also Optical cable; Optical fiber
Fiber loss, 32, 34, 39, 41, 48, 116
   See also Coupling loss
Fiber nonlinearity, 115–25
Fiber Raman amplifier, 157
Fiber splice, 39–46
Fiber-to-the-curve, 8–9
Fiber-to-the-home, 8–9
Fiber-to-the-office, 8–9
Fiber-type WDM device, 233
Field-effect transistor, 85–86
FIL. *See* Filter
Filter, 129, 135, 258, 264
Fourier transform, 35
Four-wave mixing, 115, 123–25
FPA. *See* Fabry-Perot amplifier
FP-LD. *See* Fabry-Perot laser diode

Frame synchronous bits, 195
FR bits. *See* Frame synchronous bits
Frequency-division multiple access, 19
Frequency response, 35, 54
Frequency shift keying, 54, 89, 97–99,
    102, 109, 122, 280
FSK. *See* Frequency shift keying
FTTC. *See* Fiber-to-the-curve
FTTH. *See* Fiber-to-the-home
FTTO. *See* Fiber-to-the-office
Fused-fiber devices, 65–68
Fusion splice, 43, 45, 174
FWM. *See* Four-wave mixing

Gain, 156, 161–64, 168–72, 174–75, 184, 186, 248
    See also Polarization dependence
Gaussian approximation, 31, 41, 80
Gaussian-filtered MSK, 89
GMSK. *See* Gaussian-filtered MSK
Graded-index multimode fiber, 32
Grating, 69
Group velocity dispersion, 123
GVD. *See* Group velocity dispersion

Heterodyne detection, 89, 92, 94, 97,
    101–2, 129, 284–85
High-bit-rate transmission, 41–42
High-impedance design, 134–38
Hole burning, 186
Homodyne detection, 89, 92–94, 103, 105
Hybrid star network, 9
HZ design. *See* High-impedance design

ID. *See* Identification
Identification, 19
IF band. *See* Intermediate frequency band
IM. *See* Intensity modulation
IMD. *See* Intermodulation distortion
IMP. *See* Intermodulation product
Inductance, 57
Induction peaking design, 145–47
Infrared absorption, 32–33
Inline amplifier, 159
Intensity-dependent refractive index, 122–23
Intensity modulation, 54–56, 73–79,
    109–10, 112–14, 291
Interferometer, 89, 257, 259–64
Intermediate frequency band, 88, 93
Intermodulation distortion, 291
Intermodulation product, 112, 114
IR absorption. *See* Infrared absorption

JFET. *See* Junction FET

Junction FET, 144

LAN. *See* Local-area network
Large-scale integrated circuits, 4
Laser attenuator, 211–18
Laser diode
    as semiconductor, 3, 13, 46–57, 155–57, 160–72
    in time-compression multiplexing, 197–227
    tunable, 247–53
Laser transceiver, 197, 209–211, 226
LATT. *See* Laser attenuator
LD. *See* Laser diode
LD transceivers. *See* Laser transceivers
LED. *See* Light-emitting diode
Light-emitting diode, 3
Line coding, 78–79
    See also Coded mark inversion; Nonreturn
    to zero; Return to zero
Linewidth of laser, 56–57
Local area network, 11
Low-impedance design, 144–45
Low-pass filter, 95, 97
LPF. *See* Low-pass filter
LSI. *See* Large-scale integrated circuits
LZ design. *See* Low-impedance design

Mach-Zehnder interferometer, 89, 259–62
MAN. *See* Metropolitan area network
Mass-fusion splice, 43
Material dispersion, 34–35
MCVD. *See* Modified chemical vapor deposition
Mechanical splice, 44–45
Metropolitan area network, 241
Minimum shift keying, 89
Modal dispersion, 34
Modal noise, 234
Modified chemical vapor deposition, 31
Modulation, 13, 16, 53–56, 89, 91, 94,
    97–100, 280–83
    See also Intensity modulation; Phase modulation;
    Pulse-code modulation; Self-phase
    modulation
Mono-coated fiber, 37–38
Mono-mode fiber. *See* Single-mode fiber
MQW. *See* Multiple quantum well
MSK. *See* Minimum shift keying
Multichannel systems, 109, 236–45, 278–85
Multichannel video distribution, 291–97
Multielectrode laser diode, 248–50
Multimode fiber, 32, 34–35
Multimode lasing, 53
Multiple-access network, 289, 303
Multiple quantum well, 170

Multiplexer, 4, 7, 9, 11–19, 106, 253–65
 See also Code-division multiplexing; Directional division multiplexing; Optical frequency-division multiplexing; Space-division multiplexing; Subcarrier multiplexing; Time-compression multiplexing; Wavelength-division multiplexing
MUX. *See* Multiplexer
MZ interferometer. *See* Mach-Zehnder interferometer

NDFWM. *See* Nearly degenerate four-wave mixing
Nearly degenerate four-wave mixing, 186
Networks, 9–11
Noise, 61–65, 79, 84–85, 94, 104, 112–15, 177–84, 222–23, 225, 234, 291, 293, 296–97
Nonreturn to zero, 14, 79, 132
NRZ. *See* Nonreturn to zero

O/E converter. *See* Optical/electrical converter
OFDM. *See* Optical frequency-division multiplexing
OFDMA. *See* Optical frequency-division multiple access
OH bits. *See* Overhead bits
OMI. *See* Optical modulation index
OPLL. *See* Optical phase-locked loop
Optical amplifiers
 noise and, 177–83
 outline of, 155–60, 183
 response and, 184–85
 See also Doped-fiber amplifiers; Laser diode
Optical cable, 37–41
Optical/electrical converter, 4
Optical feedback design, 142–44
Optical fiber, 29–37, 73
Optical-Fiber Time Domain Analysis, 122
Optical frequency-division multiple access, 23, 26
Optical frequency-division multiplexing, 88, 105–10, 124, 241
 multi/demultiplexer, 253–65
 passive optical network, 265–77
 simultaneous amplification, 186–89
Optical isolator, 42
Optical modulation index, 112
Optical phase-locked loop, 93
Optical receiver design
 classification of, 129–34
 comparison of, 145–50
 front-end, 134–45
 high-impedance, 134–38
 induction peaking, 145
 low-impedance, 144–45
 resonance-type, 145
 transimpedance, 138–44
Optical signal processing, 4, 299–303
Optical subscriber loops
 classification of, 8–9
 network topology, 9–11
 system outline, 3–9
Optical time domain reflectometer, 39
OTDA. *See* Optical-Fiber Time Domain Analysis
OTDR. *See* Optical time domain refectometer
Outer vapor deposition, 31
Output power, 200–204
OVD. *See* Outer vapor deposition
Overhead bits, 32–33, 195

Packet mode, 28
Passive double star network, 9, 195
Passive optical network, 9, 19–26, 265–77
Passive triple star network, 9
Pattern-dependent gain, 184
PC. *See* Polarization controller
PCM. *See* Pulse-code modulation
PD. *See* Photodiode
PDS. *See* Passive double star network
Peak detector, 150
Peltier, 78
Personick analysis, 85–86
Phase-locked loop, 93
Phase modulation, 109, 122
Phase-shift keying, 73, 89, 100, 103, 109, 122–23
Photodetection, 61, 63, 197–211, 218, 221
Photodiode, 4, 13, 26, 57–65, 219, 223
Photonic integrated circuits, 299–300
Photon lifetime, 54
Pin-photodiode receiver, 150–51, 210
Plasma effect, 248
PLL. *See* Phase-locked loop
PM. *See* Phase modulation
PN. *See* Pseudorandom noise
Poisson distribution, 61, 64
Polarization, 31, 94, 96
Polarization controller, 168, 273
Polarization dependence, 134, 168–72, 177, 205–9, 214–15, 218
PON. *See* Passive optical network
Postamplifier, 156
PR bit. *See* Preamble bit
Preamble bit, 195
Preamplifier, 159, 218–27
Propagation, 35
Pseudorandom noise, 221
PSK. *See* Phase-shift keying
PTS. *See* Passive triple star network

Pulse-code modulation, 79, 129

QAM. *See* Quadrature amplitude modulation
Quadrature amplitude modulation, 89
Quantum limit, 85, 87, 104
Quantum noise. *See* Shot noise

Raman scattering. *See* Stimulated Raman scattering
Rayleigh scattering, 39
RC circuit. *See* Resistance-capacitance circuit
Receiver. *See* Optical receiver design
Reflection, 13–14, 41–42, 45, 53, 161–63, 248, 250
Refractive index, 29, 31, 34–35, 45, 48, 56, 113, 122–23, 248, 250
Relative intensity noise, 112–13
Relaxation frequency, 54, 56
Remote terminal, 4, 9
Resistance-capacitance circuit, 135
Resonance-type design, 145–46
Responsivity, 59–60, 200–209, 214, 241
Return to zero, 79
RIN. *See* Relative intensity noise
RT. *See* Remote terminal
RZ. *See* Return to zero

Satellite communication, 20
Saturation output, 164–68, 175–77
SBS. *See* Stimulated Brillouin scattering
SC. *See* Star coupler
SCM. *See* Subcarrier multiplexing
SCMA. *See* Subcarrier multiple access
Scrambling, 132
SDM. *See* Space-division multiplexing
SDMA. *See* Space-division multiple access
Second harmonic distortion, 291
Self-phase modulation, 115, 122–23
Sellmeier equation, 261
Semiconductor laser diode amplifier, 155–57, 160–72
Sensitivity, 83–84, 104–5, 132, 140, 142, 148–49, 210, 222–27, 265, 289
Shot noise, 61, 79, 104, 114–15, 291
SI fiber. *See* Step-index fiber
Signal-to-interference ratio, 108
Signal-to-noise ratio, 82, 114–15
Silica fiber, 29–31
Single-mode fiber, 31–32, 36, 65, 234
Single-mode lasing, 53
Single-photodiode detection, 92, 94
Single star network, 9, 195
SIR. *See* Signal-to-interference ratio
SLDA. *See* Semiconductor laser diode amplifier
S/N. *See* Signal-to-noise ratio
Soliton, 4, 123

Space-division multiple access, 19, 24
Space-division multiplexing, 11–12, 16
Speckle pattern, 234–35
Splice. *See* Fiber splice
SPM. *See* Self-phase modulation
SRS. *See* Stimulated Raman scattering
SS network. *See* Single star network
Star coupler, 19, 106, 159, 240, 253, 265, 290
Step-index fiber, 30
Stimulated Brillouin scattering, 115, 119–22
Stimulated Raman scattering, 115–20
STM. *See* Synchronous transfer mode
Stokes light, 115–17
Subcarrier multiple access, 24
Subcarrier multiplexing, 11–13, 18–19, 109–15, 129, 131, 229–31, 240–45, 289–97
Synchronous demodulation, 95, 98–102
Synchronous detection, 89
Synchronous transfer mode, 26–28

Taylor's expansion series, 37
TCM. *See* Time-compression multiplexing
TDMA. *See* Time-division multiple access
Telephone office, 3, 26
TE mode. *See* Transverse electric mode
Temperature dependence, 209
Thermal noise, 79, 84–85, 134, 140, 142–43, 145, 153, 225–26
Thermistor, 78
Third harmonic distortion, 291
3R function, 4
Time-compression multiplexing, 11–13, 15, 19, 122, 195–97
    laser diodes and, 196–227
Time-division multiple access, 19, 22, 196–97
Timing extraction, 152
TM mode. *See* Transverse magnetic mode
TO. *See* Telephone office
Transfer function, 134, 140, 142
Transfer modes, 26–28
Transimpedance design, 138–44
Transmission loss, 235–36, 240
Transverse electric mode, 168
Transverse magnetic mode, 168
Traveling wave amplifier, 155–56, 158, 160, 165, 169, 171, 179–81, 183, 188, 222
Trunk systems, 4–7
Tuning, 247–54, 261–62, 264
TWA. *See* Traveling wave amplifier
TZ design. *See* Transimpedance design

Ultraviolet absorption, 32–33
Upgrading, 229–236, 274

Upstream transmission, 13, 16
UV absorption. *See* Ultraviolet absorbtion

VAD. *See* Vapor-phase axial deposition
Vapor-phase axial deposition, 31

Waveform, 85–86
Waveguide devices, 65–68, 233
Waveguide dispersion, 34–35
Wavelength demultiplexer, 268
Wavelength dependence, 60, 134, 165–66,
        204–5, 215, 218, 259–61
Wavelength-division multiple access, 23, 26
Wavelength-division multiplexing, 9, 11–12, 65,
        68–70, 88, 186–89
   multichannel systems, 236–245
   upgrading, 229–36
Wavelength-selective star coupler, 265
WD. *See* Wavelength demultiplexer
WDM. *See* Wavelength-division multiplexing
WDMA. *See also* Wavelength-division
        multiple access
Weakly guiding fibers, 31
Wentzel-Kramers-Brillouin method, 35
WKB method. *See* Wentzel-Kramers-Brillouin
        method
WSSC. *See* Wavelength-selective star coupler